Lecture Notes in Mathematics 1465

Editors:
A. Dold, Heidelberg
B. Eckmann, Zürich
F. Takens, Groningen

Subseries: Nankai Institute of Mathematics
Tianjin, P. R. China (vol. 8)

Advisers: S. S. Chern, B.-j. Jiang

Guy David

Wavelets and Singular Integrals on Curves and Surfaces

Springer-Verlag
Berlin Heidelberg New York
London Paris Tokyo
Hong Kong Barcelona
Budapest

Author

Guy David
Université Paris-Sud
Mathématiques (Bât. 425)
91405 Orsay Cédex, France

1st Edition 1991
2nd, Corrected Printing 1992

In this corrected printing the bibliography has been augmented and updated.

Mathematics Subject Classification (1991): 42B20, 42B25

ISBN 3-540-53902-6 Springer-Verlag Berlin Heidelberg New York
ISBN 0-387-53902-6 Springer-Verlag New York Berlin Heidelberg

Typesetting: Camera-ready by author/editor
Printing and binding: Druckhaus Beltz, Hemsbach/Bergstr.
2146/3140-543210 - Printed on acid-free paper

INTRODUCTION

These notes are the transcript of a series of lectures that were held in the Nankai Institute of Mathematics, in June 1988, as a part of the program on harmonic analysis.

This book consists of three parts devoted to the following topics : wavelets, Calderón-Zygmund operators, and singular integral operators on some curves and rectifiable subsets of $\mathbb{R}^n$. Our aims for the three parts are slightly different. The first part is intended to be an introduction to the theory of wavelets, and will insist mostly on the construction of various orthonormal bases of L^2. The second part is centered on a necessary and sufficient condition for the L^2-boundedness of certain singular integral operators on $\mathbb{R}^n$ (the so-called T(b)-theorem), and some of its applications. Although the result is not very recent, the techniques to prove it have not been described, up to now, in too many sources. Thanks to a recent work of R. Coifman and S. Semmes, we shall be able to give a full, conceptually very simple proof of that theorem. The final part is a survey of the author's main area of interest : questions related to the description of those k-dimensional subsets of $\mathbb{R}^n$ on which analogues of the Cauchy kernel define bounded operators on L^2. The area is quite recent, but there are already sufficiently many results, often of a fairly technical nature, to justify the description attempted in Part 3. Unfortunately, there will not be enough room for a complete description of all recent results, but we shall try to allude to most of them, and to give a precise account of the fundamental techniques involved.

This book is not intended to be exhaustive. We shall try instead to explain in details a few of the "real-variable methods" that have appeared after the books [St], [CM1] and [Jé]. The lectures in Nankai and, to a lesser extent, these notes, were prepared for an audience whose main center of interest would be close to classical analysis. This means that, in many cases, we shall avoid spending time on some of the classical results of harmonic analysis. A fair knowledge of these results, although not really required, will often help the reader understand our motivations, for instance.

The first part is about wavelets. Actually, we shall only consider orthonormal wavelets, i.e. wavelets that form orthonormal bases of $L^2(\mathbb{R}^n)$. These bases are obtained from one, or a finite number of functions ψ, by dyadic translations and dilations. For instance, the Haar system is a basis of this type, but the corresponding functions ψ are not, in this case, continuous functions. The interesting wavelets will be asked to have both a good decay at infinity and some reasonable amount of smoothness. The main advantage, from our point of view, is that the decomposition of a function f in such a basis then looks a lot like a Littlewood-Paley decomposition of f, but of course is simpler and also uniquely determined. In practical terms, this will imply that the size of the coefficients of f, as well as the type and speed of convergence of the series that gives f back, will depend in a very simple way on the smoothness and decay properties of the function f. Also, the fact that the wavelets have enough decay makes the "wavelet transform" (i.e., the mapping that sends f to its coefficients in the wavelet basis) a local transform : one does not need to know a function on the whole real line to have an idea of its coefficients near 0. This is of course an advantage over the Fourier transform (which does not have these two nice properties).

Most of Part 1 will be devoted to the construction of various orthonormal bases. A few of the properties of the wavelet transforms will also be mentioned, but we shall restrict ourselves to those which have a direct connection to harmonic analysis. This means that most of the applications of wavelets, in particular the practical ones, will be omitted. The author is aware that this is a very incomplete choice of topics, but does not think he has enough time, or competence, to treat much more material. Moreover, the interested reader will easily find all the missing information in the very good books by Y. Meyer entitled "Ondelettes et opérateurs" [My4]. Incidentally, most of Part 1 was prepared out of a preliminary version of [My4], which French-speaking readers might as well consult directly.

Part 1 begins with a small introduction to the notion of wavelets. All our constructions of wavelets will use the concept of "multiscale analysis" introduced by S. Mallat and Y. Meyer. This concept is presented in Sections 2 and 3. The orthonormal basis of $L^2(\mathbb{R})$ associated to a given multiscale analysis is constructed in Section 4. This gives, for instance, Y. Meyer's "Littlewood-Paley wavelet" $\psi \in \mathcal{S}(\mathbb{R})$, or wavelets arising from spline functions. In Section 5, we shall see how build a basis of $L^2(\mathbb{R}^n)$ using one-dimensional wavelets and a tensor product construction. It is also possible to associate an orthonormal basis of $L^2(\mathbb{R}^n)$ to a multiscale analysis even when it does not come from a tensor product ; we shall see how in Section 6. Section 7 is devoted to the construction of I. Daubechies' compactly supported wavelets. In Section 8, we shall briefly describe some of the properties of wavelets that make them so attractive (this section can be read before the previous ones). A few words will also be said about the existence of various extensions.

The second part of these notes is about singular integral operators. What we now call Calderón-Zygmund operators (see Definitions 1-1, 1-2 and 1-5) is a much wider class than the first examples that were first studied by Calderón and Zygmund in the fifties. Of course, many of the properties of the early examples remain true in the more general case : for instance, the L^p-boundedness, $1<p<\infty$, always follows from the L^2-boundedness (see Theorem 1-6). The main difference, though, is in how to get the L^2-boundedness. When dealing with convolution operators, this was quite easy, but it became much harder for operators like the commutators of

Calderón, or the Cauchy integral operator on a Lipschitz graph. We shall concentrate, in this part, on an L^2-boundedness criterion (the T(b)-theorem), and then give a few of the classical applications (essentially, those that are related to our part 3). It is interesting to note that we shall use, for this part, the same sort of techniques as in Part 1. Our proof of the T(b)-theorem, for instance, is a very nice illustration of how ideas coming from wavelets can be used with success on Calderón-Zygmund operators. This proof is due to R. Coifman and S. Semmes, and its idea is the following. When T is such that both T and its transpose send the function b to 0, one finds a suitable Riesz basis (the Haar system works when $b \equiv 1$), and then one proves that the coefficients of the matrix of T in that basis decay, sufficiently fast, away from the diagonal. The choice of a basis made of step functions simplifies the computations a great deal.

After a first section of introduction, we shall construct, in Section 2, the Riesz basis associated to a given paraaccretive function b. This basis will then be used to prove, with the help of Schur's lemma, that T is bounded in the special case when $Tb = T^t b = 0$ (Section 3). The general case will follow from this and the construction of a paraproduct (Section 4). The proof extends very nicely to spaces of homogeneous type (see the remarks of Section 5). Most of the few applications mentioned in the final section will be variants of the theorem of Coifman, McIntosh and Meyer that says that the Cauchy integral defines a bounded operator on L^2 for any Lipschitz graph. For a broader point of view, we refer to [Ch1] or [My4].

Our third part is, in a way, a continuation of the second one. Can one extend Coifman, McIntosh and Meyer's theorem to curves other than Lipschitz graphs, and then, even worse, to higher-dimensional non-smooth surfaces?

There are a few approaches to try to answer the question, but by lack of time we shall give most of our attention to only one of them (the author's favorite). The idea is the following. If $\mathcal{S}$ is a k-dimensional Lipschitz graph in $\mathbb{R}^n$, then the singular kernels we want to consider always define bounded operators on $L^2(\mathcal{S})$. This is easily deduced from Coifman, McIntosh and Meyer's result on the Cauchy kernel in one dimension. One can then try to use the standard estimates on the kernel and "good λ inequalities" to extend the L^2-boundedness results from Lipschitz graphs to more and more general sets (this is pretty much in the spirit of the pioneers of Calderón-Zygmund theory).

The systematic part of this approach will be described in Sections 2 and 3. Once this is done, one is left with a problem on the geometry of the non-smooth surface $\mathcal{S}$ (something like this : does $\mathcal{S}$ contain big pieces of Lipschitz graphs?). This problem is simple enough when $\mathcal{S}$ is a curve, and one can then give a necessary and sufficient condition, on a rectifiable curve $\mathcal{S}$, for the Cauchy integral to define a bounded operator on $L^2(\mathcal{S})$: $\mathcal{S}$ must be Ahlfors-regular (see Definition 4-1). Also, a minor modification of the argument for regular curves can be used to deduce the boundedness of the Cauchy integral on any Lipschitz graph from the case when the Lipschitz norm is small enough (i.e., from Calderón's theorem). This is done in Section 4.

In dimensions higher than one, or even for one-dimensional non-connected sets, things are not so simple. In Section 5, we shall say a few words about Garnett's counterexample to convince the reader that the geometry of the situation is much more complicated (and interesting) than for curves. We shall then give, in Section 6, three classes of surfaces on which, for various reasons, singular integrals define bounded operators. We shall mention in Section 7 a few cases when the "good λ approach" described above is successful. Section 8 contains the proof of a nice result of P. Jones concerning the existence of sets on which a given Lipschitz function has a bilipschitz restriction.

About one year after the lectures, a new Section 9 was added to the others. It concerns more recent results, and more particularly another way to deal with our problem. In [Jn2], in order to study the L^2-boundedness of the Cauchy integral operator on curves, P. Jones showed that Lipschitz graphs satisfy what he calls a geometric lemma. It was discovered later that, even in higher dimensions, the geometric lemma was, as P. Jones put it, lurking in the background. A second tool used by P. Jones in [Jn2] is a variant of the stopping-time argument often referred to as the corona construction. Most of Section 9 is concerned with recent results using those ideas. Part 3 ends with a Section 10 on further questions.

The official author of these notes would like to thank Peng Lin and Jiu-Ping Zhong, who helped preparing this manuscript. He is also very thankful to Paule Truc, from Ecole Polytechnique, and Josette Dumas, From Paris 11 (Orsay), who kindly accepted to type it.

The author is very grateful to Y. Meyer for allowing him to use, quite extensively, a preliminary version of his book [My4] for the preparation of the first part of these notes ; he is even more grateful for all the time Y. Meyer spent explaining him the little he knows about wavelets. It is also a pleasure to thank R. Coifman, P. Jones and S. Semmes for many enriching conversations, and for providing the author with the details of the various proofs that can be found in these notes.

Finally, the author wishes to thank Professor S. S. Chern, Professor M. T. Cheng and the Nankai Institute for inviting him, and his patient audience for their friendly welcome. He hopes to be able to return to China, in such a pleasant atmosphere, very soon.

The kind reader is asked not to pay too much attention to the quality of english used in these notes.

TABLE OF CONTENTS

Part I : WAVELETS

1. Introduction

As was mentioned earlier, this part is nothing more than extracts from Y. Meyer's book [My4]. Our approach here, however, will be strongly biased : I will try to take the point of view of someone interested mostly in Calderón-Zygmund operators, rather than numerical analysis. For a more natural point of view, it is strongly recommended to anyone that can read some french to report directly to "the book" [My4].

Let us say a few words about motivations. The term "wavelets " is meant to suggest "small waves", and the idea is to oppose them to the "long ones", like the sine and cosine. We wish to find ways of decomposing a function f into a series (or an integral) involving functions (say, ψ_I for $I \in \mathcal{A}$) given in advance.

For instance, we could take the $e^{ix.\xi}$, $\xi \in \mathbf{R}^n$, and use the Fourier transform to get $f(x) = \int \widehat{f}(\xi)e^{ix.\xi}d\xi$ (note that the constant $(2\pi)^{-n/2}$ was omitted ; we shall go on omitting constants like this, and we'll try to compensate by always taking an even number of Fourier transforms).

Another example, for periodic functions, is the Fourier expansion $f(x) = \sum_{k\in\mathbf{Z}^n} c_k\, e^{ik.x}$ where, this time, the sum is discrete.

We will try to find expansions like

$$f = \sum_{I\in\mathcal{A}} c_I\, \psi_I \quad \text{(where } \mathcal{A} \text{ is some set of indices)} \tag{1}$$

It will definitely be a good thing if c_I can be computed, for instance by $c_I = \int f\overline{\psi}_I$ (it will not be too bad if $c_I = \int f\widetilde{\psi}_I$, for some other set of $\widetilde{\psi}_i$'s). Of course, it will be even better if the ψ_I's are an orthonormal basis, because then the the c_I's will be unique, and also, perhaps, because in this case the c_I's are a minimal amount of information that allows to recover f.

We also want to ask the functions ψ_I to be obtained from a single one (or may be a finite family) by translations and dilations.

Decompositions like (1) have been known for quite some time. For instance, recall Calderón's reproducing formula. Let $\psi \in \mathrm{L}^1(\mathbf{R}^n)$ be real, radial and such that $\int \psi = 0$ and $\int_0^\infty |\widehat{\psi}(t\xi)|^2 \frac{dt}{t} \equiv 1$. Set $\psi_t(x) = t^{-n}\psi(x/t)$ for $t > 0$. The formula is

$$f = \int_0^\infty \psi_t * \psi_t * f\ \frac{dt}{t} \quad , \text{for } f \in L^2(\mathbf{R}^n). \tag{2}$$

Now, we can see (2) as a way to associate coefficients to f, and then reconstruct f from the coefficients : for $a > 0$ and $b \in \mathbf{R}^n$, call $\psi_{a,b} = a^{-n/2}\psi(\frac{x-b}{a})$ (the "wavelet" with "scale" a and "centered" at b). The "wavelet transform" can be defined by

(3) $$F(a,b) = \int f(x)\,\psi_{a,b}(x)dx\,,$$

and (2) can be seen as the "inversion formula"

(4) $$f(x) = \iint_{\substack{a>0 \\ b\in\mathbf{R}^n}} F(a,b)\psi_{a,b}(x)\frac{dadb}{a^{n+1}}$$

(we leave the details as an exercise).

This formula is far from being recent. The reader should be warned that, for most of the applications of wavelets , the formula (2) would be quite sufficient. However, introducing orthogonal wavelets will, in most cases, make computations much simpler and, in some cases, the fact that one has a basis will be really necessary.

Note that, with a more clever choice of ψ, one can simplify (2) a little, and in particular replace the integral by a discrete sum (see [FJ1,2,3]). Also, reproducing formulae such as (2) were introduced later (and independently) by A. Grossman (for theoretical physics reasons) and J. Morlet (for oil prospection reasons).

We shall look more specifically for functions ψ, defined in $\mathbf{R}$, such that the functions $\psi_{j,k}(x) = 2^{j/2}\psi(2^j x - k)$, $j,k \in \mathbf{Z}$ are an orthonormal basis of $L^2(\mathbf{R})$. Furthermore, we shall ask ψ to be rapidly decreasing, and reasonably smooth. Let us rapidly explain why.

An example of function ψ would be $\mathbf{1}_{[0,1/2]} - \mathbf{1}_{[1/2,1]}$, which would give the Haar basis. When we take a function f, and write it as a series $\Sigma\, c_{j,k}\psi_{j,k}$, the partial sums are never continuous (even if f is very regular), and will, in general, converge to f only in L^2-norm. If the function ψ is smooth, then the partial sums are smooth too, and it is possible to hope that if f is regular, then the partial sums will converge to f in a much smaller space than L^2. Without getting ahead of ourselves, let us mention that it is indeed the case : if ψ is smooth enough (and rapidly decreasing) and f is in the Sobolev space H^s, then partial sums will indeed converge to f for the topology of H^s ! A connected fact is that many function spaces will be characterized (again if ψ is smooth enough, and rapidly decreasing) by the fact that the coefficients $c_{j,k} = \int \overline{\psi}_{j,k}$ vanish at a certain rate when j tends to $+\infty$. We shall be a little more precise later.

Before we start constructing wavelets , let us mention a little bit of history. The first occurence of "smooth, orthogonal wavelets ", is due to O. Stromberg, in 1981 : he found, for each integer r, a function ψ of class C^r that decays exponentially at ∞, and such that the $\psi_{j,k}$'s are an orthonormal basis of $L^2(\mathbf{R})$ [Sg]. A function $\psi \in S(\mathbf{R})$ with the same property was discovered (again independently) by Y. Meyer (in 1985), and rapidly generalized to n dimensions by P.G. Lemarié [LM].

The approach that will be described here is quite different from the initial ones. It relies on the notion, introduced by S. Mallat, of "multiscale analysis".

2. Multiscale Analysis.

The following notion will help us understand the construction of the wavelets . It will also make the proofs a little longer : in some cases, we could just pull out a formula for ψ, and ask the reader to check that it works (actually, the first proofs looked a little like this to the unexperienced reader).

Let us mention, before we start, that the construction below is due to S. Mallat and Y. Meyer.

Definition 2.1.— *A multiscale analysis of $L^2(\mathbf{R}^n)$ is an increasing sequence V_j, $j \in \mathbf{Z}$, of closed subspaces of $L^2(\mathbf{R}^n)$, with the following properties :*

(5) $\bigcap_{j\in\mathbf{Z}} V_j = \{0\}$ *and* $\bigcup_{j\in\mathbf{Z}} V_j$ *is dense ;*

(6) $f(x) \in V_j$ *if and only if* $f(2x) \in V_{j+1}$ *;*

(7) $f(x) \in V_0$ *if and only if* $f(x-k) \in V_0$ *for each* $k \in \mathbf{Z}^n$ *;*

(8) There is a function $g \in V_0$ *such that the functions* $g(x-k)$, $k \in \mathbf{Z}^n$, *form a Riesz basis of* V_0.

Recall that (8) means that the $g(x-k)$ generate a dense subspace of V_0, and that there is a $C \geq 0$ such that

$$(9) \quad C^{-1}\Big(\sum_{k\in\mathbf{Z}^n} |\alpha_k|^2\Big)^{1/2} \leq \Big\|\sum_{k\in\mathbf{Z}^n} \alpha_k\, g(x-k)\Big\|_{V_0} \leq C\Big(\sum_{k\in\mathbf{Z}^n} |\alpha_k|^2\Big)^{1/2}$$

for all ℓ^2-sequences α_k (or all finite sequences α_k).

Equivalently, it means that $(a_k) \mapsto \sum_{k\in\mathbf{Z}^n} \alpha_k g(x-k)$ defines an isomorphism from $\ell^2(\mathbf{Z}^n)$ onto V_0.

Comments :

- If we know V_0, (6) tells us how to define the V_j's, and so we only need to check (5), (7), (8) and the monotonicity of the V_j's.
- The function g is not unique. Many different g's could give the same multiscale analysis.
- A trivial consequence of (6) and (7) is that $f(x) \in V_j$ if and only if $f(x-k2^{-j}) \in V_j$ for all $k \in \mathbf{Z}^n$.
- Here is a vague interpretation of the role of the V_j's. For each j, call P_j the orthogonal projection on V_j ; (5) says that $P_j f \to 0$ when $j \to -\infty$, and $P_j f \to f$ when $j \to +\infty$ (in both case, with strong convergence [Exercise]). Each j corresponds roughly to the size 2^{-j}, and the projection $P_j f$ we'll give details about f, up to the size $\simeq 2^{-j}$. So, $P_j f$ should give more and more precise details (we'll see this later, in examples).

Example 1 : splines of order r .

For this example, the dimension is $n = 1$. We define $V_0 = \{f \in \mathrm{L}^2(\mathbf{R}) : f$ is of class C^{r-1} and the restriction of f to each interval $]k, k+1[$ is a polynomial of degree $\leq r\}$. For instance, if $r = 0$ we get step functions, whereas $r = 1$ gives piecewise affine functions that are continuous. The spaces V_j are defined using (6), and the properties (5) and (7) are clear. We'll see soon that (8) is true, and even that we can pick $g = \chi * \chi \cdots * \chi$ $(r+1$ times), where χ is the characteristic function of [0,1].

Definition 2.2.— *Let $r \in \mathbf{N}$. We'll say that the multiscale analysis (V_j) is r-regular if one can choose the function g in (8) so that*

$$|\partial^\alpha g(x)| \le C_{M,\alpha}(1+|x|)^{-M} \tag{9}$$

for all $M \ge 0$ and all multiindices α such that $|\alpha| \le r$.

Note that we only ask $\partial^\alpha g$ to be in L^∞ but not necessarily continuous. If we believe that $g = \chi * \cdots * \chi$ gives a Riesz basis for V_0, then splines of order r give a M.S.A. (multiscale analysis) which is r-regular.

Example 2 :

Let us try $V_0 = \{f \in L^2(\mathbf{R}) : \widehat{f} \text{ is supported on } [-\pi,\pi]\}$ in dimension 1. It is easy to check that V_0 is stable by translations by integers and that (5) is true (with the obvious definition of V_j). Also, the V_j's are an increasing sequence. The problem with this example will be to find a suitable function g.

If we look for an orthonormal basis of V_0, we might as well use the Fourier transform, because the $\mathbf{1}_{[-\pi,\pi]}\,e^{ik\xi}$, $k \in \mathbf{Z}$, are an orthogonal basis of $\widehat{V}_0$. Applying $\mathcal{F}^{-1}$ gives a function $g(x) = \frac{\sin \pi x}{\pi x}$. But g does not decay rapidly at ∞. Furthermore, no other choice of $g \in V_0$ would both give a Riesz basis, and enough decay at ∞ (we'll leave this as an exercise). So we'll have to exclude this example from the authorized MSA's.

Note, by the way, that since $g(0) = 1$ and $g(k) = 0$ for $k \neq 0$, writing $f \in V_0$ as $\sum_k \lambda_k g(x-k)$ is quite easy : $\lambda_k = f(k)$.

Example 3 :

This one is a smoothed-up version of Example 2 ; this is the example that will give Y. Meyer's basis. Let $\theta(\xi)$ be a function in $\mathcal{D}(\mathbf{R}) = C_c^\infty(\mathbf{R})$, with the properties

(10) $0 \le \theta(\xi) \le 1$ everywhere ;

(11) θ is even ;

(12) $\theta(\xi) = 1$ for $\xi \in [\frac{-2\pi}{3}, \frac{2\pi}{3}]$;

(13) $\theta(\xi) = 0$ out of $[\frac{-4\pi}{3}, \frac{4\pi}{3}]$ and

(14) $\theta^2(\xi) + \theta^2(2\pi - \xi) = 1$ for $0 \le \xi \le 2\pi$ (see the picture).

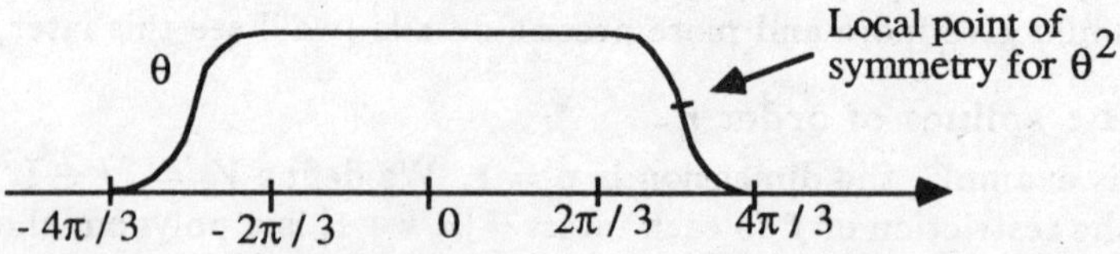

Note that if g is defined by $\widehat{g} = \theta$, then the $g(x-k)$, $k \in \mathbf{Z}$, are an orthogonal system (we leave this as an exercise, with the following hint : try on the Fourier transform side !). Define V_0 to be the closed space spanned by the $g(x-k)$, and V_j, $j \in \mathbf{Z}$, by the rule (6). To see that we have a MSA, we still need to check (5) and the fact that the V_j's are increasing.

Since

$$V_0 = \{f \in L^2 : f(x) = \sum \alpha_k \, g(x-k) \quad \text{for some sequence} \quad (\alpha_k) \in \ell^2(\mathbf{Z})\},$$

we see that

$$\widehat{V_0} = \{\widehat{f} \in L^2 : \widehat{f} = \sum \alpha_k \, e^{ik\xi}\theta(\xi) \quad \text{for some sequence} \quad (\alpha_k) \in \ell^2(\mathbf{Z})\}$$

$$= \{\widehat{f} \in L^2 : \widehat{f} = m\theta\,, \quad \text{where} \quad m \quad \text{is} \quad 2\pi-\text{periodic} \quad \text{and in} \quad L^2([-\pi,\pi])\}.$$

Since the $g(x-k)$ are (a constant times) an orthonormal basis, we even get that $\|f\|_{L^2} = c\|m\|_{L^2([-\pi,\pi])}$, where m is the 2π-periodic function such that $\widehat{f} = m\theta$. Now (5) follows because $\widehat{V_j} = \{\widehat{f} \in L^2 : \widehat{f}(\xi) = m(2^{-j}\xi)\theta(2^{-j}\xi)$ for some 2π-periodic $m \in L^2_{\text{loc}}\}$. The fact that $V_j \subset V_{j+1}$ also follows from that description of V_j and the fact that $\theta(\xi) = m(\xi/2)\theta(\xi/2)$ for a suitable m. So the V_j's are a MSA, which is as regular as we want because $g \in \mathcal{S}(\mathbf{R})$.

Example 4. (tensor product) :

Suppose the V_j's are a MSA on $L^2(\mathbf{R})$, and let us build a MSA on $L^2(\mathbf{R}^2)$. Let g be the function defined in (8), and consider the function $g(x)g(y)$, defined on $\mathbf{R}^2$. One easily checks that the functions $g(x-k)g(y-\ell)$, $(k,\ell) \in \mathbf{Z}^2$, are a Riesz system. Denote by $V_0 \otimes V_0$ the closed subspace of $L^2(\mathbf{R}^2)$ spanned by this Riesz system :

$V_0 \otimes V_0 = \{f \in L^2(\mathbf{R}^2) : f(x,y) = \sum\sum \alpha_{k,\ell}\, g(x-k)g(y-\ell)$ for a sequence $\alpha_{k,\ell} \in \ell^2(\mathbf{Z} \times \mathbf{Z})\}$.

Calling $V_j \otimes V_j$ the subspace of $L^2(\mathbf{R}^2)$ defined from $V_0 \otimes V_0$ by the rule (6), we see that the $V_j \otimes V_j$'s are an increasing sequence of closed subspaces, that satisfy (5)-(8), and which defines a MSA of $L^2(\mathbf{R}^2)$ of the same regularity as the V_j's.

Exercise : define a MSA for $L^2(\mathbf{R}^n)$.

Let us now go back to one-dimensional spline functions of order r, and prove what we said earlier about the function $g = \mathcal{X} * \cdots * \mathcal{X}$ (r+1 times). Remember that V_0 is composed of the L^2-functions that are of class C^{r-1}, and whose restriction to any interval $[k, k+1[$ is a polynomial of degree $\leq r$. If $f \in V_0$, then $\left(\frac{d}{dx}\right)^{r+1} f = \sum\limits_k c_k\delta_k$, where δ_k is the Dirac mass at $k \in \mathbf{Z}$. Since f, restricted to $[k-1, k[$ and $[k, k+1[$, is a polynomial of degree $\leq r$, and f is of class C^{r-1}, one easily check that $c_k^2 \leq \frac{1}{C}\int_{k-1}^{k+1} |f|^2$ and so $\sum\limits_k |c_k|^2 < +\infty$. Then, the function $q(\xi) = (i\xi)^{r+1}\widehat{f}(\xi)$ is a 2π-periodic function, with $\|q\|_{L^2([-\pi,\pi])} \leq C\|f\|_2$. Next, define $m(\xi) = q(\xi)(1-e^{i\xi})^{-r-1}$: this is also a 2π-periodic function, and it is in L^2 because, near 0, $m(\xi) \sim \widehat{f}(\xi)$. As a consequence, we can write $\widehat{f}(\xi) = m(\xi)\left(\frac{1-e^{i\xi}}{i\xi}\right)^{r+1}$ for some 2π-periodic $m \in L^2([0, 2\pi[)$.

Conversely, if $\hat{f}(\xi)$ is of the form $m(\xi)\left(\frac{1-e^{i\xi}}{i\xi}\right)^{r+1}$ for some 2π-periodic $m \in L^2_{\text{loc}}$, then $f \in L^2$ and $(\frac{d}{dx})^{r+1} f = \sum c_k \delta_k$ for a sequence $(c_k) \in \ell^2(\mathbf{Z})$, and it is easily checked that $f \in V_0$ (exercise). So $\widehat{V}_0 = \{m(\xi)\left(\frac{1-e^{i\xi}}{i\xi}\right)^{r+1} : m$ is 2π-periodic and in $L^2([0,2\pi])\}$.

Note that this looks a lot like our description of V_0 in the Examples 2 and 3, except that this time θ does not decrease rapidly at infinity. In fact, the rapidity of the decay depends on r, and this will account for the regularity of g and of the MSA.

From our description of V_0, we see that we can choose $g = \chi * \cdots * \chi$ ($r+1$ times) to define a Riesz basis of V_0 (g is indeed the inverse Fourier transform of $\left(\frac{1-e^{i\xi}}{i\xi}\right)^{r+1}$).

Remark. When r is odd, one can find $g \in V_0$ such that $g(0) = 1$ and $g(k) = 0$ for $k \neq 0$; when r is even, it is impossible to find such a g (and so the computation of the coefficients of $f \in V_0$ in the basis defined by g is not so easy). For more details, see the book [My4].

Example : When $r = 1$, $g(x) = \chi * \chi(x)$ is 0 for $x < 0$ or $x > 2$, x for $0 \le x \le 1$, and $2 - x$ for $1 \le x \le 2$.

Exercise. Compute $g(x)$ for $r = 3$ (the solution is in [My4]).

3. The function φ .

We wish to replace the function g of (8) by a function $\varphi \in V_0$ such that the $\varphi(x-k)$, $k \in \mathbf{Z}^n$, will be an orthonormal basis of V_0.

One systematic way of doing this would be to use Gram matrices (let us call that Gram's orthogonalization process). The idea is to apply to the basis $(g(x-k))$ the inverse of the square root of the positive matrix of scalar products, so that the result is an orthonormal basis. The advantage of this process is that, countrary to the Gram-Schmidt orthogonalization process, this one does not put any order on the elements of the basis. As a consequence, if one starts with a basis which is invariant under the action of a group (like here, with the translations of $\mathbf{Z}^n$), the result is also invariant under the action of the group.

In the case that we want to study, one can use the Fourier transform to get precise formulae, and this gives a slightly faster construction (one can check that the construction given below, however, coincides with the one obtained from Gram's process). In more general cases, the Fourier transform cannot be used any longer, and Gram's process becomes the providential tool.

To find our function $\varphi \in V_0$, let us write $\varphi(x) = \sum_k c_k\, g(x+k)$ for some ℓ^2-sequence (c_k). Then $\widehat{\varphi}(\xi) = \sum c_k e^{ik\xi}\widehat{g}(\xi) = m(\xi)\widehat{g}(\xi)$ for some 2π-periodic (in all directions) function $m(\xi)$. One even has $\|\varphi\|_2 \sim \|m\|_{L^2([0,2\pi]^n)}$.

Now let us compute scalar products. If h, ℓ are in $\mathbf{Z}^n$,

$$\int \varphi(x+h)\overline{\varphi}(x+h+\ell)dx = \int (\widehat{g}(\xi)m(\xi)e^{ih\xi})(\widehat{g}(\xi)m(\xi)e^{i(h+\ell)\xi})^{-} d\xi$$

$$= \int |\widehat{g}(\xi)m(\xi)|^2\, e^{-i\ell\xi} d\xi\,.$$

Calling $G(\xi) = \sum_{j\in\mathbf{Z}^n} |\widehat{g}(\xi+2\pi j)|^2$, we get

$$\text{(15)}\qquad \int \varphi(x+h)\overline{\varphi}(x+h+\ell)dx = \int_{[0,2\pi]^n} G(\xi)\,|m(\xi)|^2\, e^{-i\ell\xi} d\xi\,.$$

Taking $h = \ell = 0$, we get $\|\varphi\|_2^2 = \int_{[0,2\pi]^n} G(\xi)|m(\xi)|^2 d\xi$. Since this formula is valid for any function $\varphi \in V_0$, and since $\|\varphi\|^2 \simeq \|m\|^2$, we see that there is a constant $c > 0$ such $G(\xi) \geq c$ almost everywhere.

Let us pause in our search for φ to study G a little more.

Lemma 3.1.— *If the MSA (V_j) is 0-regular, $G(\xi)$ is of class C^∞.*

To prove this, we'll only use the fact that $G(\xi) = \sum_{j\in\mathbf{Z}} |\widehat{g}(\xi+2\pi j)|^2$, where g is a bounded function that decreases rapidly at infinity. More precisely, for any $M > 0$, $\int |g(x)^2|(1+(x))^{2M} dx < +\infty$, and so $\widehat{g}$ is in the Sobolev space $H^M(\mathbf{R}^n)$.

Let $\theta \in C_c^\infty(\mathbf{R}^n)$ be such that $\theta(x) \equiv 1$ on $B(0, 20n)$. For each $M > 0$, the sequence $k \mapsto \|\widehat{g}(\xi)\theta(\xi - k\pi)\|_{H^M}$ is in $\ell^2(\mathbf{Z}^n)$ [exercise ; use Leibnitz'rule and the fact that if functions have disjoint supports and their sum is in L^2, then their L^2-norms form a ℓ^2-sequence]. By Sobolev's embedding's theorem, we see that for all $M' > 0$, the sequence $k \to \|\widehat{g}(\xi)\theta(\xi - k\pi)\|_{C^{M'}}$ is also in $\ell^2(\mathbf{Z}^n)$.

This is enough to conclude that the series $\sum_{j\in\mathbf{Z}^n} |\widehat{g}(\xi+2\pi j)|^2$ converges uniformly on each compact. The same thing would be true if $\widehat{g}$ were replaced by any of its derivatives. Therefore, G is in $C^\infty(\mathbf{R}^n)$, and we proved Lemma 3.1.

Since we know that $G \geq c > 0$ a.e., it follows that $G(\xi)^{-1/2}$ is a well-defined, positive and C^∞ function [the expert reader will note the similarity with Gram's process].

As (15) would suggest, let us try to define φ by

$$\text{(16)}\qquad \widehat{\varphi}(\xi) = m(\xi)\widehat{g}(\xi)\,, \quad \text{where} \quad m(\xi) = G(\xi)^{-1/2}\,.$$

Lemma 3.2.— *The function φ defined by (16) satisfies the regularity conditions (9) with the same r as g, and $\varphi(x-k)$, $k \in \mathbf{Z}^n$, is an orthonormal basis of V_0.*

Let us first prove (9) for φ. Since $m(\xi)$ is smooth, it can be written as $\sum_{k\in\mathbf{Z}^n} \alpha_k e^{ik\xi}$, where the α_k's are decaying as fast as we wish at infinity. It is then an easy exercise to check that $\varphi(x) = \sum \alpha_k\, g(x+k)$, and its derivatives of order $\leq r$, have the same decay as the derivatives of order $\leq r$ of g.

The fact that the $\varphi(x-k)$ are an orthonormal system is not a surprise, because

$$\int \varphi(x+h)\overline{\varphi}(x+h+\ell)dx = \int_{[0,2\pi]^n} e^{-i\ell\xi}\, dx = (2\pi)^n \delta_{\ell=0}$$

by (15) and (16).

If we didn't normalize the wrong way, this means that the $\varphi(x-k)$ are an orthonormal basis of some subspace $\widetilde{V}_0$ of V_0. To check that $\widetilde{V}_0 = V_0$, note that $\widehat{V}_0$ is composed of all products $n(\xi)\widehat{g}(\xi)$, where n is 2π-periodic in all directions and locally in L^2. The same computation shows that $(\widetilde{V}_0)^\wedge$ is composed of all products $n'(\xi)\widehat{\varphi}(\xi) = n'(\xi)m(\xi)\widehat{g}(\xi)$, where n' is also 2π-periodic and locally in L^2. Since m is C^∞, 2π-periodic and invertible, $\widetilde{V}_0 = V_0$, and we proved the lemma.

Remark about uniqueness.

If we wanted φ to define an orthonormal system, then we had to choose $G(\xi)|m(\xi)|^2$ so that all its Fourier coefficients (except the 0th) would be zero. So $|m(\xi)|^2 = G(\xi)^{-1}$ was necessary. On the other hand, multiplying $m(\xi)$ by any 2π-periodic, C^∞ function of modulus 1 would have given a function φ with exactly the same properties. Finally, one could always try to multiply $m(\xi)$ by unimodular functions that are not smooth, which could give the stupid result of finding a φ which does not have enough decay ! Taking for $m(\xi)$ the positive square root of $G(\xi)^{-1}$ has the advantage of being natural, and of giving the same function φ as Gram's orthogonalization process.

4. Construction of ψ in dimension 1.

We are now ready to construct the wavelets in dimension 1. Remember that we want to find a function ψ such that the $\psi_{j,k}(x) = 2^{j/2}\psi(2^j x - k)$, for $j \in \mathbf{Z}$ and $k \in \mathbf{Z}$, form an orthonormal basis of $L^2(\mathbf{R})$. The idea is to find a ψ such that, for each j, the $2^{j/2}\psi(2^j x - k)$, $k \in \mathbf{Z}$, form an orthonormal basis of the orthogonal complement of V_0 in V_1.

Call W_0 the orthogonal complement of V_0 in V_1, and let us state what we want as a theorem.

Theorem 4.1.— *Let (V_j) be a r-regular multiscale analysis of $L^2(\mathbf{R})$. There is a function $\psi \in V_1$, satisfying (9), and such that the $\psi(x-k)$, $k \in \mathbf{Z}$, form an orthonormal basis of W_0.*

For this function ψ, the $\psi_{j,k}(x) = 2^{j/2}\psi(2^j x - k)$, $(j,k) \in \mathbf{Z}^2$, form an orthonormal basis of $L^2(\mathbf{R})$.

Let us first check that the last statement follows from the first one. Using the rule (6), we see that, for each j, the $\psi_{j,k}$, $k \in \mathbf{Z}$, form an orthonormal basis of W_j, the orthogonal complement of V_j in V_{j+1}. We now use condition (5), which implies that $L^2(\mathbf{R})$ is the orthogonal sum of all the W_j's, $j \in \mathbf{Z}$, to see that the $\psi_{j,k}$'s , $(j,k) \in \mathbf{Z}^2$, are a basis of L^2.

Let us now look for the function ψ. Since we prefer working with functions of V_0, we will rather look for a function ψ' such that the $\psi'(x-2k)$, $k \in \mathbf{Z}$, form an orthonormal basis of W_{-1}.

First, note that $\varphi(\frac{x}{2}) \in V_{-1} \subset V_0$, and thus one can find a sequence $\alpha_k \in \ell^2(\mathbf{Z})$ such that $\frac{1}{2}\varphi(\frac{x}{2}) = \sum \alpha_k \varphi(x+k)$. Moreover, since $\alpha_k = \frac{1}{2}\int \varphi(\frac{x}{2})\overline{\varphi}(x+k)$, the sequence α_k is rapidly decreasing. Applying a Fourier transform, we get

$$\widehat{\varphi}(2\xi) = m_0(\xi)\,\widehat{\varphi}(\xi)\,, \tag{17}$$

where $m_0(\xi) = \sum \alpha_k\, e^{ik\xi}$ is 2π-periodic and C^∞.

Lemma 4.2.— $|m_0(\xi)|^2 + |m_0(\xi+\pi)|^2 \equiv 1.$

Remember that, by definition of φ,

$$\sum_k |\widehat{\varphi}(\xi + 2k\pi)|^2 = \sum_k |m(\xi)|^2 |\widehat{g}(\xi+2k\pi)|^2 = |m(\xi)|^2 G(\xi) = 1\,,$$

and therefore, $\sum_k |\widehat{\varphi}(2\xi+2k\pi)|^2 = 1$, too. Using (17),

$$\sum_k |m_0(\xi+k\pi)|^2|\widehat{\varphi}(\xi+k\pi)|^2 = 1$$

We now split the sum and use the fact that m_0 is 2π-periodic, to obtain

$$|m_0(\xi)|^2\sum_\ell |\widehat{\varphi}(\xi+2\ell\pi)|^2 + |m_0(\xi+\pi)|^2\sum_\ell |\widehat{\varphi}(\xi+\pi+2\ell\pi)|^2 = 1,$$

and so, $|m_0(\xi)|^2 + |m_0(\xi+\pi)|^2 = 1$, as promised.

From the fact that the $2^{-\frac{1}{2}}\varphi(\frac{x}{2}-k)$ form an orthonormal basis of V_{-1}, we see that

$$\begin{aligned}\widehat{V}_{-1} &= \{m(2\xi)\widehat{\varphi}(2\xi) \ : \ m \quad \text{is} \quad 2\pi-\text{periodic} \quad \text{and locally in} \quad L^2\} \\ &= \{m(2\xi)m_0(\xi)\widehat{\varphi}(\xi) \ : \ m \quad \text{is} \quad 2\pi-\text{periodic} \quad \text{and locally in} \quad L^2\}.\end{aligned}$$

We can use this description of $\widehat{V}_{-1}$, together with the fact that for each $f \in V_0$, $\widehat{f}$ can be written as $\widehat{f}(\xi) = m(\xi)\widehat{\varphi}(\xi)$ for a 2π-periodic $m(\xi)$ such that $\|f\| = \|m\|_{L^2([0,2\pi[)}$ (so that the map $f \to m$ is unitary), to find $\widehat{W}_{-1}$.

Let us determine which 2π-periodic functions $\ell(\xi)$ are orthogonal to all the $m(2\xi)m_0(\xi)$, where m is 2π-periodic on $[0, 2\pi]$. The condition is

$$\int_0^\pi m(2\xi)\{m_0(\xi)\overline{\ell(\xi)} + m_0(\xi+\pi)\overline{\ell(\xi+\pi)}\}d\xi = 0,$$

and so $m_0(\xi)\overline{\ell(\xi)} + m_0(\xi+\pi)\overline{\ell(\xi+\pi)} = 0$ a.e. on $[0,\pi]$. By Lemma 4.2, the vector with coordinates $m_0(\xi)$ and $m_0(\xi+\pi)$ is always of length 1, and so the orthogonality condition is equivalent to

$$\begin{pmatrix}\ell(\xi)\\ \ell(\xi+\pi)\end{pmatrix} = \lambda(\xi)e^{-i\xi}\begin{pmatrix}\overline{m_0}(\xi+\pi)\\ -\overline{m_0}(\xi)\end{pmatrix} \quad \text{a.e. on} \quad [0,\pi], \tag{18}$$

for some function $\lambda(\xi)$ [we added the $e^{-i\xi}$ for our later convenience].

Note that $\|\ell(\xi)\|_{L^2([0,2\pi])} = \|\lambda(\xi)\|_{L^2([0,\pi])}$ by Lemma 4.2. Also, the first line of (18) is $\ell(\xi) = \lambda(\xi)e^{-i\xi}\overline{m_0}(\xi+\pi)$, and the second line is $\ell(\xi+\pi) = \lambda(\xi)e^{-i(\xi+\pi)}\overline{m_0}(\xi+2\pi)$, so that (18) is equivalent to

$$\ell(\xi) = \lambda(\xi)e^{-i\xi}\overline{m_0}(\xi+\pi) \quad \text{a.e}, \tag{19}$$

for some π-periodic function λ.

The space $\widehat{W}_{-1}$ is thus composed of all functions $\ell(\xi)\widehat{\varphi}(\xi)$, where ℓ is as in (19). The equality between norms even allows us to get an orthonormal basis for $\widehat{W}_{-1}$ from a basis of $L^2([0,\pi])$: we can take the functions $\lambda(\xi) = \sqrt{2}e^{2ik\xi}$, $k \in \mathbf{Z}$, get the functions

$$\ell(\xi)\widehat{\varphi}(\xi) = \sqrt{2}e^{-i\xi}\overline{m_0}(\xi+\pi)\widehat{\varphi}(\xi)e^{2ik\xi}, \ k \in \mathbf{Z},$$

in $\widehat{W}_{-1}$, and then an orthonormal basis of W_{-1} is given by the $\psi'(x-2k)$, where $\widehat{\psi}'(\xi) = \sqrt{2}e^{-i\xi}\overline{m_0}(\xi+\pi)\widehat{\varphi}(\xi)$. By a dilation, we see that a basis of W_0 is the $\psi(x-k)$, $k \in \mathbf{Z}$, where

$$\widehat{\psi}(\xi) = e^{-i\xi/2}m_0(\frac{\xi}{2}+\pi)\widehat{\varphi}(\frac{\xi}{2}). \tag{20}$$

As before, since m_0 is C^∞ and 2π-periodic, we see that ψ satisfies (9) because φ does and ψ is a linear combination of translates of φ with coefficients that decay rapidly at ∞.

This finishes the proof of Theorem 4.1. Let us end this section with a few examples.

Example 1 : (Spline functions of order 0)

In this case, we can take $\varphi = g = \mathbf{1}_{[0,1]}$. To compute m_0, note that

$$\frac{1}{2}\varphi(\frac{x}{2}) = \frac{1}{2}\varphi(x) + \frac{1}{2}\varphi(x-1),$$

so that

$$\widehat{\varphi}(2\xi) = \frac{1}{2}(1+e^{-i\xi})\widehat{\varphi}(\xi), \quad \text{and} \quad m_0(\xi) = \frac{1}{2}(1+e^{-i\xi}).$$

Next,

$$\widehat{\psi}(\xi) = e^{-i\xi/2}\frac{1}{2}(1-e^{i\xi/2})\widehat{\varphi}(\xi/2),$$

and so

$$\psi(x) = \varphi(2x-1) - \varphi(2x) = \mathbf{1}_{[1/2,1]} - \mathbf{1}_{[0,1/2]}.$$

We miraculously recovered the Haar system !

Exercise : Compute what happens for splines of order 2, or look in [My4].

Example 2 : (Y. Meyer's wavelets).

Let us consider the multiscale analysis of Example 3, Section 2. In this case, we can take $\varphi = g = \check{\theta}$, and so

$$\widehat{\psi}(\xi) = e^{-i\xi/2}m_0(\frac{\xi}{2}+\pi)\theta(\frac{\xi}{2})$$

where m_0 is such that $\theta(2\xi) = m_0(\xi)\theta(\xi)$. Note that $\theta(2\xi)$ is supported on $[-\frac{2\pi}{3}, \frac{2\pi}{3}]$, at the place where $\theta(\xi) \equiv 1$. So $m_0(\xi) = \theta(2\xi)$ for $-\pi \le \xi \le \pi$ (and is defined everywhere else by 2π-periodicity), and so

$$\widehat{\psi}(\xi) = e^{-i\xi/2}\theta(\xi/2)[\theta(\xi+2\pi) + \theta(\xi-2\pi)].$$

Note that when one of the two functions $\theta(\xi/2)$ and $[\theta(\xi+2\pi)+\theta(\xi-2\pi)]$ vary, then the other one is constant and equal to 0 or 1 (see fig.1). Thus, $\widehat{\psi}(\xi)$ looks like the product by $e^{-i\xi/2}$ of the function θ_1 of fig.2.

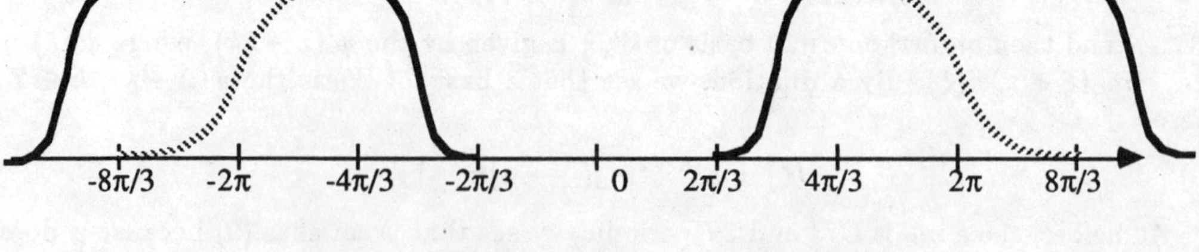

Figure 1: $\theta(\xi/2)$ and $[\theta(\xi+2\pi) + \theta(\xi-2\pi)]$.

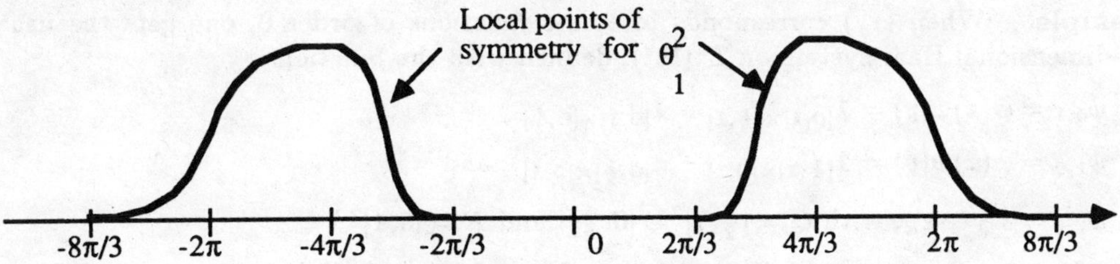

Figure 2 : $\theta_1(\xi) = e^{i\xi/2}\widehat{\psi}(\xi)$.

The graph is also (locally) invariant by $(x,y) \to (\frac{\pm 8\pi}{3} - x, y)$.

Remark : Calling Δ_j the orthogonal projection on W_j, note that Δ_j is a projection of Littlewood-Paley type. This is certainly a justification of our earlier statement that "V_j contains the part of functions with frequencies lower than 2^j".

5. Wavelets in higher dimensions : tensor product.

Remember that if (V_j) is a MSA on $L^2(\mathbf{R})$, we have a MSA of $L^2(\mathbf{R}^2)$, denoted by $V_j \otimes V_j$, which is obtained as in Example 4 of §2. Let us build a basis of $L^2(\mathbf{R}^2)$, using the $\psi_{j,k}$'s given by Theorem 4.1 (applied to (V_j)). The following construction is due to P.G. Lemarié.

From $V_1 = V_0 \perp W_0$ (the symbol means "orthogonal sum") we get $V_1 \otimes V_1 = (V_0 \otimes V_0) \perp (V_0 \otimes W_0) \perp (W_0 \otimes V_0) \perp (V_0 \otimes W_0)$ (exercise). Now, the $\varphi(x-k)\psi(y-\ell)$, $(k,\ell) \in \mathbf{Z}^2$, are an orthonormal basis of $V_0 \otimes W_0$, and the $\psi(x-k)\varphi(y-\ell)$ (respectively, $\psi(x-k)\psi(y-\ell)$ are an orthonormal basis of $W_0 \otimes V_0$ (respectively, $W_0 \times W_0$). Thus, calling $\psi_{0,1}(x,y) = \varphi(x)\psi(y)$, $\psi_{1,0}(x,y) = \psi(x)\varphi(y)$ and $\psi_{1,1}(x,y) = \psi(x)\psi(y)$, we see that an orthonormal basis of $L^2(\mathbf{R}^2)$ is given by the $2^j\psi_\varepsilon(2^jx - k_1, 2^jy - k_2)$ where $\varepsilon \in \{(0,1),(1,0)(1,1)\}$ and $j, k_1, k_2 \in \mathbf{Z}$.

The case of $L^2(\mathbf{R}^n)$ is similar, so we'll just give the formula, and leave the verifications as an easy exercise.

Call E the set of sequences $(\varepsilon_1, \ldots, \varepsilon_n) \in \{0,1\}^n$ that are different from $(0,0\ldots0)$. For $\varepsilon \in E$, set

$$\psi_\varepsilon(x) = \psi_\varepsilon(x_1, \ldots, x_n) = \psi^{\varepsilon_1}(x_1)\psi^{\varepsilon_1}(x_2)\ldots\psi^{\varepsilon_n}(x_n)$$

where $\psi^0 = \varphi$ and $\psi^1 = \psi$. An orthonormal basis of $L^2(\mathbf{R}^n)$ is obtained by taking all the functions

$$2^{nj/2}\psi_\varepsilon(2^jx - k) \quad \text{for} \quad \varepsilon \in E \quad , \quad j \in \mathbf{Z}, \quad \text{and} \quad k \in \mathbf{Z}^n .$$

Example. When (V_j) corresponds to spline functions of order 0, one gets the usual two-dimensional Haar system on $L^2(\mathbf{R}^2)$, defined with the functions

$$\psi_{0,1} = \varphi(x)\psi(y) = \mathbf{1}_{[0,1]\times[\frac{1}{2},1]} - \mathbf{1}_{[0,1]\times[0,\frac{1}{2}]} \, ,$$

$$\psi_{1,0} = \varphi(x)\psi(y) = \mathbf{1}_{[\frac{1}{2},1]\times[0,1]} - \mathbf{1}_{[0,\frac{1}{2}]\times[0,1]} \, , \text{ and}$$

$$\psi_{1,1} = \mathbf{1}_Q - \mathbf{1}_R, \text{ with } Q = [\tfrac{1}{2},1]^2 \cup [0,\tfrac{1}{2}]^2 \text{ and } R = [0,1]^2 \setminus Q.$$

Exercise : Can one take the tensor product of bases $L^2(\mathbf{R})$ coming from different MSA's ?

6. Other wavelets in higher dimensions.

Suppose we have a MSA on $L^2(\mathbf{R}^n)$ which is not obtained by taking the tensor product of one-dimensional MSA's. We still want to find an orthonormal basis of $L^2(\mathbf{R}^n)$ adapted to our MSA. As in the tensor product case, we'll need 2^n-1 different functions ψ, from which we'll build a basis of W_j (we still call (V_j) our MSA, and W_j the orthogonal complement of V_j in V_{j+1}) and then a basis of $L^2(\mathbf{R}^n)$ by the usual scheme. As far as I know, the following construction is due to K. Gröchenig and Y. Meyer.

Theorem 6.1.— *Let (V_j) be a r-regular multiscale analysis of $L^2(\mathbf{R}^n)$. One can find 2^n-1 functions ψ_ε, $\varepsilon \in E = \{0,1\}^n \setminus \{0\}$, such that each ψ_ε satisfies (9), and the $\psi_\varepsilon(x-k)$, $\varepsilon \in E$ and $k \in \mathbf{Z}^n$, are an orthonormal basis of W_0. Then, the $2^{nj/2}\,\psi_\varepsilon(2^j x - k)$, $\varepsilon \in E$, $j \in \mathbf{Z}$ and $k \in \mathbf{Z}^n$ are an orthonormal basis of $L^2(\mathbf{R}^n)$.*

The last statement follows from the first one exactly like in dimension 1. Finding the ψ_ε's is a little more complicated than in dimension 1, but the spirit is the same.

Remark : Our argument will use the action of a group of translations, and the existence of a subgroup of index 2^n ; it could be generalized to other situations with other groups (see [My4]). Although we shall not use that generality, it is a good thing to have in mind that we are using group actions.

As before, we can write $\widehat{\varphi}(2\xi) = m_0(\xi)\widehat{\varphi}(\xi)$ for some 2π-periodic function m_0 and, since $m_0(\xi) = \sum \alpha_k e^{ik\xi}$ with a sequence $\alpha_k = \frac{1}{2}\int \varphi(\frac{x}{2})\overline{\varphi}(x+k)dx$ which is rapidly decreasing, m_0 is C^∞. Since the functions ψ_ε, $\varepsilon \in E$, that we are looking for are in V_1, we can write

$$\widehat{\psi}_\varepsilon(2\xi) = m_\varepsilon(\xi)\widehat{\varphi}(\xi) \tag{21}$$

for some function $m_\varepsilon(\xi)$ which is 2π-periodic in all directions.

To simplify the notations, we shall add $\psi_{(0,0,\ldots,0)} = \varphi$ to our set of functions, and $m_{(0,0,\ldots,0)} = m_0$, so that (21) holds for all $\varepsilon \in E' = \{0,1\}^n$.

Let us cut the series giving each m_ε into 2^n pieces, and write

$$m_\varepsilon = \sum_{\eta \in E'} e^{i\eta\xi} m_{\varepsilon,\eta}(2\xi)\,, \tag{22}$$

where, for each class η in $\mathbf{Z}^n/(2\mathbf{Z})^n$, we group all terms of the Fourier series of m_ε in the class defined by η. The functions $m_{\varepsilon,\eta}$ are 2π-periodic.

Our problem is to find functions $m_{\varepsilon,\eta}$, $\varepsilon \in E$ and $\eta \in E'$, in such a way that the $\psi_\varepsilon(x-k)$, $\varepsilon \in E'$ and $k \in \mathbf{Z}^n$, form an orthonormal basis of V_1.

Let us denote by $U(\xi)$ the $2^n \times 2^n$ matrix with rows and columns indexed by $\varepsilon \in E'$ and $\eta \in E'$, and general term $m_{\varepsilon,\eta}(\xi)$.

Lemma 6.2.— *The functions $\psi_\varepsilon(x-k)$, $\varepsilon \in E'$ and $k \in \mathbf{Z}^n$, form an orthonormal basis of V_1 if and only if the matrix $2^{n/2}U(\xi)$ is unitary fort almost every ξ.*

To prove the lemma, let us first compute $\int \psi_{\varepsilon'}(x+k')\overline{\psi_\varepsilon(x+k)}dx$ for $\varepsilon, \varepsilon' \in E'$ and $k, k' \in \mathbf{Z}^n$. As before, it is convenient to compute scalar products on the Fourier transform side. Using $\|\{m\widehat{\varphi}\}^\vee\| = \|m\|_{L^2([0,2\pi]^n)}$, then replacing ξ by 2ξ, and then using the fact that m is 2π-periodic, we get

$$\begin{aligned}
\int \psi_{\varepsilon'}(x+k')\overline{\psi}_\varepsilon(x+k)dx &= c\int \widehat{\psi}_{\varepsilon'}(\xi)(\widehat{\psi}_\varepsilon)^-(\xi)e^{i(k'-k)\xi}d\xi \\
&= 2^n c\int \widehat{\psi}_{\varepsilon'}(2\xi)(\widehat{\psi}_\varepsilon)^-(2\xi)e^{2i(k'-k)\xi}d\xi \\
&= 2^n c\int_{[0,2\pi]^n} m_{\varepsilon'}(\xi)\widehat{\varphi}(\xi)\overline{m_\varepsilon(\xi)\widehat{\varphi}(\xi)}e^{2i(k'-k)\xi}d\xi \quad \text{by (21)} \\
&= 2^n c\int_{[0,2\pi]^n} m_{\varepsilon'}(\xi)\overline{m_\varepsilon(\xi)}e^{2i(k'-k)\xi}d\xi \\
&= c\int_{[0,4\pi]^n} m_{\varepsilon'}(\xi/2)\overline{m_\varepsilon(\xi/2)}e^{i(k'-k)\xi}d\xi
\end{aligned}$$

$$= c\int_{[0,4\pi]^n} e^{i(k'-k)\xi}\sum_{\eta'}\sum_{\eta} e^{i\eta'\xi/2}m_{\varepsilon',\eta'}(\xi)e^{-i\eta\xi/2}\overline{m_{\varepsilon,\eta}(\xi)}d\xi \quad \text{(by 22)}.$$

Notice that all the $m_{\varepsilon,\eta}$ and $m_{\varepsilon',\eta'}$ are 2π-periodic in all variables. If we perform the integral with respect to ξ_n first, and split $[0,4\pi]$ in two equal parts, we will get opposite results unless $\eta_n = \eta'_n$. The same thing would be true if we integrated with respect to any other variable, and so the only terms left correspond to $\eta = \eta'$:

$$\begin{aligned}
\int \psi_{\varepsilon'}(x+k')\overline{\psi_\varepsilon(x+k)}dx &= 2^n c\int_{[0,2\pi]^n} e^{i(k'-k)\xi}\sum_{\eta\in E'} m_{\varepsilon',\eta}(\xi)\overline{m_{\varepsilon,\eta}(\xi)}d\xi \\
&= c\int_{[0,2\pi]^n} e^{i(k'-k)\xi}\langle v'(\xi), v(\xi)\rangle d\xi\,,
\end{aligned}$$

where $v(\xi)$ is the vector with coordinates $2^{n/2}m_{\varepsilon,\eta}(\xi)$, $\eta \in E'$, $v'(\xi)$ is the vector with coordinates $2^{n/2}m_{\varepsilon',\eta}(\xi)$, and $\langle\ \ \rangle$ is the usual scalar product in $\mathbf{C}^{2^n}$.

If the $\psi_\varepsilon(x-k)$ are an orthonormal basis , our computation shows that the Fourier series of $\langle v'(\xi), v(\xi)\rangle$ must be zero when $\varepsilon' \neq \varepsilon$ and so $\langle v'(\xi), v(\xi)\rangle = 0$ for almost every ξ. Also, for $\varepsilon' = \varepsilon$, the Fourier series of $\langle v'(\xi), v(\xi)\rangle$ is the same as the Fourier series of 1, so $\|v(\xi)\| = 1$ for almost every ξ. So the matrix $2^{n/2}U(\xi)$ is unitary almost everywhere, and the condition of the lemma is necessary.

Conversely, if $2^{n/2}U(\xi)$ is unitary a.e., the computation above shows that the $\psi_\varepsilon(x-k)$ are an orthonormal basis of some subspace of V_1, and we only need to check that it is the whole V_1.

Let us proceed by contradiction, and suppose that some $f \in V_1$ is orthogonal to all the $\psi_\varepsilon(x-k)$. Let $m(\xi)$ be the 2π-periodic function such that $\widehat{f}(2\xi) = m(\xi)\widehat{\varphi}(\xi)$. Let us decompose $m(\xi)$ like in (22), to obtain $m(\xi) = \sum_{\eta \in E'} e^{i\eta\xi} m'_\eta(2\xi)$ for 2^n functions m'_η that are 2π-periodic. Following the same computation as above, we get

$$\int f(x)\overline{\psi_\varepsilon}(x+k)dx = \int_{[0,2\pi]^n} e^{-ik\xi}\langle w(\xi), v(\xi)\rangle d\xi\,,$$

where $w(\xi)$ is the vector with coordinates $m'_\eta(\xi)$, and $v(\xi)$ is as before. Since this is zero for all $\varepsilon \in E'$ and $k \in \mathbf{Z}^n$, we get $\langle w(\xi), v(\xi)\rangle = 0$ for all $\varepsilon \in E'$ and a.e. ξ. Since $2^{n/2}U(\xi)$ is unitary, the $v(\xi)$ span the whole space, and $w(\xi) = 0$ a.e. This of course implies that $f = 0$, and proves Lemma 6.2.

Now we see that finding the ψ_ε's (for $\varepsilon \in E$) reduces to the following problem : we are given the first row of a $2^n \times 2^n$ matrix, and we want to find $2^n - 1$ other rows, so that the resulting matrix be unitary. Also, we want to do this for each ξ, but in a continuous way.

First note that our first row is of length 1 for almost all ξ, because the $\varphi(x-k)$, $k \in \mathbf{Z}^n$, are an orthonormal basis of V_0 (just apply the computation of Lemma 6.2 with $\varepsilon = \varepsilon' = 0$). A natural attempt to complete the matrix would be to associate, to each vector of the unit sphere of $\mathbf{R}^{2^n}$ (or $\mathbf{R}^{2^{n+1}}$ in the complex case), a unitary matrix which has this vector as its first row. The trouble is that such a mapping from vectors to matrices, if we also require it to be continuous, does not always exist !

To avoid this difficulty, we shall use a trick, due to K. Gröchenig. The idea is that our first row will not take all possible values in the unit sphere, and so we only need to define the mapping from vectors to matrices in a proper subset of the unit sphere (and then the topological obstructions disappear).

Note that $\xi \mapsto (2^{n/2}m_{0,\eta}(\xi))_{\eta \in E'}$ is a C^∞ mapping from $\mathbf{R}^n$ to a sphere of dimension $2^n - 1 > n$ (because $n \geq 2$) in the real valued-case, and $2^{n+1} - 1 > n$ in the complex case. Consequently, the image by $(2^{n/2}m_{0,\eta})_{\eta \in E'}$ of $[0,2\pi]^n$ has zero surface measure and, since it is compact, there is a small ball, centered on the sphere, which does not meet the image $[2^{n/2}(m_{0,\eta})_{\eta \in E'}]([0,2\pi]^n)$. We shall use the following lemma.

Lemma 6.3.— *Let S be the unit ball of $\mathbf{R}^{2^n}$ (respectively $\mathbf{C}^{2^n}$), and B a small ball centered on S. There is a mapping F, defined and C^∞ on $S \setminus B$, with values in the set of unitary, $2^n \times 2^n$ matrices with real (respectively complex) coefficients, and such that for all $x \in S \setminus B$, x is the first row of $F(x)$.*

Let us first see how to deduce Theorem 6.1 from this lemma. Let us take, for our matrix $((2^{n/2}m_{\varepsilon,\eta}))$, the function $F((2^{n/2}m_{0,\eta}))$ given by the lemma [where B is a ball that does meet $(2^{n/2}m_{0,\eta}([0,2\pi]^n)) = (2^{n/2}m_{0,\eta}(\mathbf{R}^n))$]. Then all the $m_{\varepsilon,\eta}$ are C^∞, 2π-periodic in all directions, and the matrix $2^{n/2}U(\xi)$ of Lemma 6.2 is unitary for all ξ. It follows that the functions ψ_ε, $\varepsilon \in E'$ defined by (21) and (22) give an orthonormal basis of V_1, and so the $\psi_\varepsilon(x-k)$, $\varepsilon \in E$ and $k \in \mathbf{Z}^n$, are an orthonormal basis of W_0. Also, each $m_\varepsilon(\xi)$ is C^∞, and so the functions ψ_ε defined by (21) satisfy (9) because φ does (we have seen the argument a few times already). Thus Theorem 6.1 follows from Lemma 6.3.

Let us prove the lemma in the complex-valued case (in the real case, one would just remove the bars). Call $q = 2^n - 1$ and let $(z_1, \ldots, z_q, z_{q+1})$ denote the coordinates of a vector $z \in S \subset \mathbf{C}^{q+1}$. We can safely assume that B is centered at $(0, \ldots, 0, 1)$.

Our first step will be to find q vectors $w_1, \ldots, w_q$, so that $z, w_1, \ldots, w_q$ form a basis of $\mathbf{C}^{q+1}$. Let us take for w_j the $(j+1)^{st}$ column vector of the matrix

$$A(z) = \begin{pmatrix} z_1 & \alpha & 0 & \cdots \\ z_2 & 0 & \alpha & \cdots \\ \cdots & \cdots & \cdots & \cdots \\ z_q & 0 & \cdots & \alpha \\ z_{q+1} & -\bar{z}_1 & \cdots & -\bar{z}_q \end{pmatrix}, \tag{23}$$

where $\alpha > 0$ is chosen as a function of the size of B. We just have to show that $\det(A(z)) \neq 0$ for all $z \notin B$, if α is small enough.

Exercise : Prove that $\det(A(z)) = (-1)^{q+1}\{\alpha^{q-1}(|z_1|^2 + \cdots + |z_q|^2) - \alpha^q z_{q+1}\}$.

Let us check that if $z \notin B$ and α is small enough, $\det A(z) \neq 0$. If this was not true, one would have $\alpha\, z_{q+1} = |z_1|^2 + \cdots + |z_q|^2$, and so $z_{q+1} > 0$ and also, since $|z_{q+1}| \leq 1$, $|1 - z_{q+1}|^2 = |z_1|^2 + \cdots + |z_q|^2 \leq \alpha$. This is impossible since z is not too close to $(0, 0, \ldots, 1)$. So, if α is small enough, the vectors $z, w_1, \ldots, w_q$ form a basis of $\mathbf{C}^{q+1}$.

Let us now apply the Gram-Schmidt orthonormalization process, starting with z so as not to change it. We get a matrix $F(z)$,which has z as its first row, is unitary, and also depends in a C^∞ manner on z, because $A(z)$ does (in the orthonormalization process, one never divides by 0, or takes the square root of 0). This proves the lemma.

7. Compactly supported wavelets .

We still postpone the description of the various properties of the various wavelets that we constructed (the interested reader can go directly to the relevant section), and construct some more.

Let us come back to one dimension, and try to construct compactly supported wavelets. The first construction that was similar to the one we are going to present was due to Tchamitchian [Tc1,2]. He constructed two functions θ and ψ, with compact support and prescribed regularity, such that the $\theta_{j,k}$ and the $\psi_{j,k}$ (with the notations of Sections 1 and 4) are two (dual) Riesz bases of $L^2(\mathbf{R})$, and such that every $f \in L^2(\mathbf{R})$ can be written

$$f = \sum c_{j,k}\, \theta_{j,k} \quad \text{with} \quad c_{j,k} = \int f(x)\overline{\psi_{j,k}(x)}\,dx\,. \tag{24}$$

The following, more precise theorem is due to I. Daubechies [Db1].

Theorem 7.1.— *There is a constant $C \geq 0$ such that, for each integer $r \geq 0$, there is a multiscale analysis of $L^2(\mathbf{R})$, of regularity r, such that the functions φ and ψ of Section 3 and Section 4 are compactly supported, with supports in $[-Cr, Cr]$.*

Note that finding a $\psi \in C_c^\infty$ is impossible (exercise : it would follow from the fact that the $\psi_{j,k}$ form an orthonormal basis that all moments of ψ are zero, which is impossible if ψ is supported in $[-M, M]$ by density of the polynomials).

The fact that ψ has compact support is a nice property, because it makes the "wavelet transform" completely local (instead of almost local). One can hope that this property will be useful for practical problems (for instance involving time !).

The proof we are going to see is not the original one. As usual, it was communicated to me by Y. Meyer, and it involves ideas of S. Mallat, Y. Meyer, J-P. Kahane and Y. Katznelson. The construction is also similar to Tchamitchian's work mentioned above.

The idea is the following. If (V_j) is a MSA of $L^2(\mathbf{R})$, we have seen in Section 4 that there is a 2π-periodic, C^∞ function $m_0(\xi)$, such that $\widehat{\varphi}(2\xi) = m_0(\xi)\widehat{\varphi}(\xi)$ (this is (17)). We shall start from a function m_0, and try to build a MSA with it. The point is that the size of the supports of ψ and φ is easier to control from m_0.

So, let us give ourselves a function m_0 with the following properties :

$$(25) \qquad \begin{cases} m_0(\xi) \quad \text{is} \quad C^\infty \text{ and } 2\pi-\text{periodic}, \\ |m_0(\xi)|^2 + |m_0(\xi+\pi)|^2 = 1 \quad \text{for all} \quad \xi \\ m_0(0) = 1 \,. \end{cases}$$

We have seen in Lemma 4.2 that the first conditions are necessary if we want m_0 to arise from (17) for a MSA. It would not be too hard to check that the third condition is necessary too (because of (17), one only needs to check that $\widehat{\varphi}(0) = 1$). Since we are only interested in the converse, we will leave the verification to the reader.

If we want to build a MSA from m_0, we will need a function φ, and here is a good candidate :

$$(26) \qquad \widehat{\varphi}(\xi) = \prod_{j=1}^{\infty} m_0(2^{-j}\xi) \,.$$

Note that the infinite product converges because $m_0(0) = 1$; indeed, $m_0(2^{-j}\xi) = 1 + O(2^{-j})$, so that $\sum_j \operatorname{Log} m_0(2^{-j}\xi)$ converges uniformly on compacta, and so $\widehat{\varphi}(\xi)$ is well-defined. We still want to check that $\varphi(x-k)$, $k \in \mathbf{Z}$, is an orthonormal system, and that its closed span V_0 is part of a MSA. Finally, we shall have to choose m_0 so that the resulting φ is regular enough, and compactly supported.

If we want φ to be compactly supported, we'll have to choose for m_0 a finite trigonometric sum, because

$$m_0(\xi) = \sum \alpha_k e^{ik\xi}, \quad \text{where} \quad \alpha_k = \frac{1}{2}\int \varphi(\frac{x}{2})\overline{\varphi(x+k)}dx \quad \text{is also compactly supported.}$$

On the other hand, suppose m_0 is a (finite) trigonometric sum, and let us show that φ is compactly supported. By (26), φ is the (infinite) convolution product of the $[2^j \check{m}_0(2^j x)]$, which are supported in $[-c2^{-j}, c2^{-j}]$, so that the support of φ is contained in the sum of the supports, i.e. $[-2c, 2c]$. [*Exercise* : check the argument, and in particular make sure it has a sense distributionwise.] Finally, if φ is compactly supported and m_0 is a finite trigonometric sum, ψ is also compactly supported because of (20). So we'll choose for $m_0(\xi)$ a trigonometric polynomial.

Let us come back to the problem of defining a MSA from a function $m_0(\xi)$ defining (25).

Lemma 7.2.— *If φ is defined by (26) for a function m_0 satisfying (25), then $\|\varphi\|_2 \leq 1$.*

We intend to prove later that the norm is exactly 1, but let us start with this. Call

$$I_N = \int_{-2^N\pi}^{2^N\pi} |\Pi_N(\xi)|^2 d\xi, \quad \text{where} \quad \Pi_N(\xi) = \prod_{j=1}^{N} m_0(2^{-j}\xi).$$

Let us compute I_N by induction. Since Π_N is $2^{N+1}\pi$-periodic,

$$I_N = \int_0^{2^{N+1}\pi} |\Pi_N(\xi)|^2 d\xi = \int_0^{2^N\pi} \cdots + \int_{2^N\pi}^{2^{N+1}\pi} \cdots.$$

In the second integral, we write $\xi = 2^N\pi + u$ and get

$$\begin{aligned} I_N &= \int_0^{2^N\pi} |\Pi_{N-1}(\xi)|^2 \left(|m_0(\frac{\xi}{2^N})|^2 + |m_0(\frac{\xi}{2^N} + \pi)|^2 \right) d\xi \\ &= \int_0^{2^N\pi} |\Pi_{N-1}(\xi)|^2 d\xi = I_{N-1} = \cdots = I_1 \\ &= \int_{-2\pi}^{2\pi} |m_0(\frac{\xi}{2})|^2 d\xi = 2\int_{-\pi}^{\pi} |m_0(\xi)|^2 d\xi \\ &= 2\int_{-\pi}^{0} (|m_0(\xi)|^2 + |m_0(\xi+\pi)|^2) d\xi = 2\pi. \end{aligned}$$

Since $|m_0(\xi)| \leq 1$, $|\widehat{\varphi}(\xi)| \leq |\Pi_N(\xi)|$, and so

$$\int_{-2^N\pi}^{2^N\pi} |\widehat{\varphi}(\xi)|^2 d\xi \leq I_N = 2\pi;$$

letting $N \to +\infty$, we get $\int_{-\infty}^{\infty} |\widehat{\varphi}(x)|^2 dx \leq 2\pi$, and so $\int_{-\infty}^{\infty} |\varphi(x)|^2 dx \leq 1$.

At this stage, we need to be a little more careful, because of the following two examples :

1) if $m_0(\xi) = \frac{1+e^{i\xi}}{2}$, then $\widehat{\varphi}(\xi) = e^{i\xi/2}\frac{\sin \xi/2}{\xi/2}$, and $\varphi = \mathbf{1}_{[-1,0]}$ (we leave the computations as an exercise, but you can use Example 1 of Section 4 if you are lazy). In this case, one obtains Haar's basis, and everything works smoothly.

2) if $m_0(\xi) = \frac{1+e^{3i\xi}}{2}$, (25) is still satisfied, but now $\varphi = \frac{1}{3}\mathbf{1}_{[-3,0]}$ (change variables in the computations of case 1)). The $\varphi(x-k)$ are no longer an orthonormal basis , so something went wrong !

Fortunately, the mishappening of case 2) is rather easy to prevent : we shall see that it is enough to ask m_0 not to take the value 0 between $\frac{-\pi}{2}$ and $\frac{\pi}{2}$.

Lemma 7.3.— *If m_0 is a trigonometric polynomial satisfying (25), and if $m_0(\xi) \neq 0$ for $\xi \in [\frac{-\pi}{2}, \frac{\pi}{2}]$, then the $\varphi(x-k)$, $k \in \mathbf{Z}$, are an orthonormal sequence.*

Note that $\varphi \in L^2(\mathbf{R})$ because of Lemma 7.2, and is compactly supported because m_0 is a trigonometric polynomial, so $\widehat{\varphi}$ is in all the Sobolev spaces. Let $G(\xi) = \sum_{-\infty}^{\infty} |\widehat{\varphi}(\xi+2k\pi)|^2$. This is the same function as in Section 3, except that now the dimension is 1 and $\varphi = g$. The proof of Lemma 3.1 still shows that G is C^∞, and the computation giving (15) (with $m = 1$) shows that the $\varphi(x-k)$, $k \in \mathbf{Z}$, are an orthonormal system if and only if $G(\xi) \equiv 1$.

Let us first check that $G(0) = 1$. Note that $\widehat{\varphi}(2\ell\pi) = 0$ if ℓ is a nonzero integer because if $\ell = 2^p q$ for some odd integer q, then $m_0(q\pi)$ shows up in the product giving $\widehat{\varphi}(2\ell\pi)$, but $m_0(q\pi) = m_0(\pi) = 0$ since $m_0(0) = 1$ and $|m_0(0)|^2 + |m_0(\pi)|^2 = 1$. It follows that $G(0) = 1$, as promised.

To prove Lemma 7.3, we only need to show that G is a constant. To do so, we shall use an argument of Tchamitchian.

Call $\alpha(\xi) = |m_0(\xi)|^2$, so that $0 \leq \alpha(\xi) \leq 1$, $\alpha(\xi) + \alpha(\xi+\pi) \equiv 1$, and $|\widehat{\varphi}(\xi)|^2 = \prod_{j=1}^{\infty} \alpha(2^{-j}\xi)$.

Let us first prove that

$$G(2\xi) = \alpha(\xi)\, G(\xi) + \alpha(\xi+\pi)\, G(\xi+\pi) \tag{27}$$

(so $G(2\xi)$ is a barycentre of $G(\xi)$ and $G(\xi+\pi)$). Write

$$\begin{aligned} G(2\xi) &= \sum_k |\widehat{\varphi}(2\xi+2k\pi)|^2 = \sum_k |m_0(\xi+k\pi)|^2 |\widehat{\varphi}(\xi+k\pi)|^2 \\ &= \sum_{k\,\text{even}} \cdots + \sum_{k\,\text{odd}} \cdots \\ &= |m_0(\xi)|^2 \sum |\widehat{\varphi}(\xi+2k\pi)|^2 + |m_0(\xi+\pi)|^2 \sum |\widehat{\varphi}(\xi+\pi+2k\pi)|^2 \\ &= \alpha(\xi)G(\xi) + \alpha(\xi+\pi)G(\xi+\pi)\,, \quad \text{as promised.} \end{aligned}$$

Call m the minimum and M the maximum of G on $\mathbf{R}$ (i.e., on $[-\pi,\pi]$)). Let $\xi_0 \in [-\pi,\pi]$, such that $G(\xi_0) = m$, and apply (27) to $\xi = \xi_0/2$. Since $\alpha(\xi) > 0$ by hypothesis ($\xi \in [\frac{-\pi}{2}, \frac{\pi}{2}]$), we get $G(\xi) = m$, too. Applying this argument to $\xi_0/2$, and then $\xi_0/4$, and so on, we get $G(\xi_0/2^n) = m$ for all n and, since G is continuous, $G(0) = m$.

We can also use the argument with a ξ_0 such that $G(\xi_0) = M$, and obtain $G(1) = M$. Consequently, G is a constant, and Lemma 7.3 is established.

We'll need the following lemma.

Lemma 7.4.— *Given $r > 0$, one can find a finite trigonometric sum $m_0(\xi)$, satisfying the hypotheses of Lemma 7.3, and such that the function φ has r bounded derivatives.*

Of course, since φ is compactly supported, it will also satisfy (9).

Let us momentarily admit the lemma, and see how we get a multiscale analysis from m_0.

Call V_0 the closed span of the $\varphi(x-k)$, $k \in \mathbf{Z}$. Because of Lemma 7.3, the $\varphi(x-k)$ are an orthonormal system, so that (8) is automatically satisfied. Define V_j, $j \in \mathbf{Z}$, by the rule (6). To prove that the V_j's are an increasing sequence, we only need to check that $V_{-1} \subset V_0$, which is true because $\frac{1}{2}\varphi(\frac{x}{2}) \in V_0$ since $\widehat{\varphi}(2\xi) = m_0(\xi)\widehat{\varphi}(\xi)$ by definition of φ. We still need to check (5).

Call E_j the orthogonal projection from L^2 onto V_j. We need to show that for each $f \in L^2$, $E_j f \to f$ as $j \to +\infty$, and $E_j f \to 0$ as $j \to -\infty$. We can compute the kernel function of E_j, because

$$E_j f = \sum_k \langle f, \varphi_{j,k} \rangle \varphi_{j,k}$$

with the usual notations. The kernel of E_j is

$$E_j(x,y) = 2^j E_0(2^j x, 2^j y)\,, \quad \text{where} \quad E_0(x,y) = \sum_k \varphi(x-k)\overline{\varphi(y-k)}\,.$$

Since φ is bounded and compactly supported, $|E_0(x,y)| \le C$, and also $E_0(x,y) = 0$ if $|x-y| > 2 \operatorname{diam}(\operatorname{supp}\varphi)$. Next,

$$\int E_0(x,y)dy = \int \sum_k \varphi(x-k)\overline{\varphi(y-k)}dy = \sum_k \varphi(x-k)$$

because $\widehat{\varphi}(0) = 0$. Since $\sum_k \varphi(x-k)$ is periodic, and

$$\int_0^1 \left\{ \sum_k \varphi(x-k) \right\} e^{-2i\pi \ell x} dx = \int_{\infty}^{\infty} \varphi(x) e^{-2i\pi \ell x} dx = \widehat{\varphi}(2\pi\ell) = \delta_{\ell=0}\,,$$

we get

$$\int E_0(x,y)dy = \sum_k \varphi(x-k) \equiv 1\,.$$

With these properties of E_0, it is easy to show that $E_j f \to f$ when $j \to +\infty$ and $E_j f \to 0$ as $j \to -\infty$ whenever $f \in C_c^\infty$, say. Since the E_j's are uniformly bounded (again by the properties of E_0, or simply because they are projections), one gets

$$E_j f \to \begin{cases} f \\ 0 \end{cases} \quad \text{as} \quad j \to \begin{cases} +\infty \\ -\infty \end{cases} \quad \text{for all} \quad f \in L^2\,,$$

and so the V_j's are a multiscale analysis.

Since the $\varphi(x-k)$, $k \in \mathbf{Z}$, are an orthonormal basis of V_0, φ is really the function of Section 3, and so the MSA (V_j) is r-regular if m_0 was chosen with the help of Lemma 7.4. Furthermore, if m_0 can be written $\sum_{T}^{T} \alpha_k e^{ik\xi}$ for some T, then the supports of the functions φ and ψ are contained in $[-CT, CT]$ for some $C \ge 0$.

So we see that we'll have established the theorem if we can prove Lemma 7.4, with a trigonometric polynomial m_0 of degree $\le Cr$.

We shall deduce the fact that φ has r bounded derivatives from the simpler inequality

(28) $$|\widehat{\varphi}(\xi)| \le C(1+|\xi|)^{-s} \quad \text{with, say,} \quad s = r+1\,,$$

and Sobolev's embedding theorem.

The following Lemma will help us. It is due to F. Riesz, and its proof is left to the reader as an exercise.

Lemma 7.5.— *Let* $g(\xi) = \sum_{-T}^{T} \gamma_k e^{ik\xi}$ *be a trigonometric polynomial which is non-negative on* $\mathbf{R}$. *Then there is a trigonometric polynomial of the form*

$$m_0(\xi) = \sum_{-T}^{T} \alpha_k e^{ik\xi} \quad \textit{such that} \quad |m_0(\xi)|^2 = g(\xi)\,.$$

Taking this lemma into account, we only have to find a non-negative trigonometric polynomial $g(\xi) = \sum_{-T}^{T} \gamma_k e^{ik\xi}$, with $T \leq Cr$, with the following properties :

(29) $$g(\xi) + g(\xi+\pi) = 1 \quad \text{for all} \quad \xi, \quad \text{and} \quad g(0) = 0\,;$$

(30) $$g(\xi) > 0 \quad \text{on} \quad [\frac{-\pi}{2}, \frac{\pi}{2}], \quad \text{and}$$

(31) $$\prod_{j=1}^{\infty} g(2^{-j}\xi) \leq C(1+|\xi|)^{-2s}$$

(condition (29) and (30) will give the hypotheses of Lemma 7.3, and (31) ensures (28)).

By the way, we shall obtain a g with real coefficients and, looking more closely at the proof of Riesz's lemma, one sees that one can get a m_0 with real coefficients α_k (we shall not need that precision).

Let us try to motivate the definition of g that follows : if we did not have to select a trigonometric polynomial, and were just interested in (31), we could chose

$$g(\xi) = \sum_k \mathbf{1}_{[\frac{-\pi}{2}+2k\pi, \frac{\pi}{2}+2k\pi]}\,.$$

The infinite product of (31) would then be supported in $[-\pi, \pi]$ because if $|\xi| > \pi$, then one of the $\xi/2^j$ falls in $]\frac{+\pi}{2}, \frac{3\pi}{2}[\cup]\frac{-3\pi}{2}, \frac{-\pi}{2}[$ and the product is zero. One can see the definition of g below as an attempt to imitate this example, and choose g as large as possible near 1 and as small as possible near π, so that the infinite product will have more chances to drop sharply.

To define g, select a large $k \in \mathbf{N}$, and call

$$c_k = \left\{ \int_0^{\pi} (\sin t)^{2k+1}\, dt \right\}^{-1}\,.$$

We shall define g by

(32) $$g(\xi) = 1 - c_k \int_0^{\xi} (\sin t)^{2k+1}\, dt\,.$$

Note that g is a trigonometric polynomial of degree $\leq 2k+1$, that $0 \leq g(\xi) \leq 1$ everywhere and $g(\xi) > 0$ on $]-\pi, \pi[$ by our choice of c_k, and that $g(\xi) + g(\xi+\pi) = 1$. So we'll just have to check that

$$G(\xi) = \prod_{j=1}^{\infty} g(2^{-j}\xi) \quad \text{satisfies} \quad (31) \quad \text{for some} \quad s \geq k/C\,.$$

Lemma 7.6.— *We have* $c_k \leq C\sqrt{k}$.

Choose a small $a > 0$, and write

$$\frac{1}{c_k} \geq \int_{\frac{\pi}{2}-a/\sqrt{k}}^{\frac{\pi}{2}+a/\sqrt{k}} (\sin t)^{2k+1} dt \geq \frac{2a}{\sqrt{k}} \left[\sin\left(\frac{\pi}{2} + \frac{a}{\sqrt{k}}\right)\right]^{2k+1}$$

$$= \frac{2a}{\sqrt{k}} \left[\cos \frac{a}{\sqrt{k}}\right]^{2k+1} .$$

Since

$$(2k+1) \operatorname{Log}\left(\cos \frac{a}{\sqrt{k}}\right) \sim (2k+1)\frac{a^2}{k} \sim 2a^2 ,$$

we obtain $c_k^{-1} \geq (C\sqrt{k})^{-1}$, and so the lemma is true.

Lemma 7.7.— *If k is large enough,*

(33) $$g(t) \leq \left(\frac{4}{5}\right)^k \quad \textit{for} \quad \frac{2\pi}{3} \leq t \leq \frac{4\pi}{3}, \quad \textit{and}$$

(34) $$g(t) \leq |t-\pi|^{2k+2} \quad \textit{for} \quad 0 \leq t \leq 2\pi .$$

To prove (34), we write $g(\xi) = c_k \int_\xi^\pi (\sin t)^{2k+1} dt$ for $\xi \in [0, 2\pi]$, and, since $|\sin u| \leq |\pi - u|$, we get

$$g(\xi) \leq c_k |\xi - \pi|^{2k+2}/2k+2 \leq |\xi - \pi|^{2k+2}$$

if k is large enough (because of Lemma 7.6).

To prove (33), we write

$$g(\xi) \leq c_k \int_\xi^\pi (\sin \frac{2\pi}{3})^{2k+1} du \leq c\sqrt{k} (\frac{\sqrt{3}}{2})^{2k+1} \leq (\frac{4}{5})^k$$

if k is large enough.

Let us summarize in a lemma a few computations that we shall do later.

Lemma 7.8.— *Given $0 < \delta < 1$ and $C \geq 0$, there exists $\alpha > 0$ such that the following is true. Let $f(t)$ be continuous, 1-periodic, such that $0 \leq f(t) \leq 1$ and*

(35) $$0 \leq f(t) \leq \delta \quad \textit{for} \quad \frac{1}{3} \leq t \leq \frac{2}{3} \quad \textit{and}$$

(36) $$0 \leq f(t) \leq C|t - \frac{1}{2}| \quad \textit{for} \quad 0 \leq t \leq 1 .$$

Then $F_j(t) = f(t)\, f(2t) \ldots f(2^{j-1}t)$ satisfies

(37) $$\sup_{\frac{1}{4} \leq t \leq \frac{1}{2}} F_j(t) \leq 2^{-\alpha j} \quad \textit{for} \quad j \geq 1 .$$

Before proving this lemma, let us show how it implies that $G(\xi)$ satisfies (31). A change of variables shows that

$$\sup_{\frac{1}{4}\le t\le\frac{1}{2}} F_j(t) = \sup_{2^{j-2}\le t\le 2^{j-1}} f(\frac{t}{2})\dots f(\frac{t}{2^j})$$

and so (37) implies that

$$\prod_{\ell=1}^{\infty} f(2^{-\ell}t) \le 2^{-\alpha j} \quad \text{for} \quad t\in[2^{j-2},2^{j-1}] \quad \text{and} \quad j\ge 1\,. \tag{38}$$

Let us take $f(t) = [g(2\pi t)]^{1/(2k+2)}$, so that (33) and (34) imply (35) and (36) with suitable values of δ and C. Given $\xi\in[2^{j-1}\pi, 2^j\pi]$, we can apply (38) to $t=\xi/2\pi$ and obtain

$$G(\xi) = \prod_{\ell=1}^{\infty}[f(2^{-\ell}t)]^{2k+2} \le 2^{-\alpha(2k+2)j} \le C(1+|\xi|)^{-2\alpha(k+1)}\,.$$

A similar estimate would be true for $\xi\in[-2^{j-1},-2^{j-2}]$, and so we proved (31) with $s=\alpha(k+1)$. As we said before, this proves that φ has $\alpha(k+1)-1$ bounded derivations, and so, if we choose $k\ge\alpha^{-1}r+\alpha^{-1}$ we get the function $m_0(\xi)$ required for Lemma 7.4, therefore proving the theorem.

Thus we only need to prove Lemma 7.8. To do so, it is more convenient to consider the function $h(t) = f(t)\,f(2t)$. If we can prove that $h(t)\,h(2t)\dots h(2^{j-1}t)\le 2^{-\alpha j}$, this will imply that $f(t)\,f(2t)^2\dots f(2^{j-1}t)^2\; f(2^jt)\le 2^{-\alpha j}$, and then $f(t)\dots f(2^jt)\le 2^{-\alpha j/2}$, which will imply (37) with a different value of α.

The function h is a little better than f, because it satisfies

$$0\le h(t)\le\delta \quad \text{for} \quad \frac{1}{6}\le t\le\frac{5}{6} \quad \text{and} \tag{39}$$

$$0\le h(t)\le C|t-\frac{1}{2}| \quad \text{for} \quad 0\le t\le 1\,. \tag{40}$$

We want to show that $H(t) = h(t)\; h(2t)\dots h(2^{j-1}t)$ satisfies

$$H(t)\le 2^{-\alpha j} \quad \text{for} \quad \frac{1}{4}\le t<\frac{1}{2} \tag{41}$$

(the case of $t=\frac{1}{2}$ is not a problem because $h(\frac{1}{2})=0$).

Write t in base 2 : $t = 0\cdot\frac{1}{2}+1\cdot\frac{1}{4}+\alpha_3\frac{1}{8}+\cdots$ or, in short, $t=(0,1,\alpha_3\cdots)$. Note that if $q\ge 0$, $h(2^qt) = h(\alpha_{q+1},\alpha_{q+2},\cdots)$ by periodicity.

Given t, call $F\subset\{0,1,\cdots(j-1)\}$ the set of q's such that $\alpha_{q+1}\ne\alpha_{q+2}$. Let us see why $h(2^qt)$ should be small when $q\in F$. If $q\in F$, 2^qt is of the form $(1,0,\cdots)$ or $(0,1,\cdots)$ modulo 1, and so it is between $\frac{1}{2}$ and $\frac{1}{2}+\frac{1}{4}<\frac{5}{6}$ in the first case, and between $\frac{1}{4}$ and $\frac{1}{2}$ in the second case. In both cases, (39) yields

$$h(2^qt)\le\delta\,. \tag{42}$$

If there are enough $q \in F$, this estimate will be good enough. Otherwise, we will have to use (40) a few times. Let $q \in F$. If $2^q t$ is of the form $(1,0\ldots)$ modulo 1, call ℓ_q the number of consecutive zeros that are just after the 1 ; if $2^q t$ is of the form $(0,1\ldots)$ modulo 1, let ℓ_q be the number of consecutive 1's after the initial zero. Let us give ourselves a large constant $A \geq 0$ (we shall decide on its value in a moment). If $\ell_q \geq A$, we shall use (40). Indeed, $2^q t$ looks like $(1,0,\ldots 0,\ldots)$ (with ℓ_q zeros) or $(0,1,1\ldots 1,\ldots)$ (with ℓ_q 1's) modulo 1, and so it is at distance $\leq 2^{-\ell_q}$ from $\frac{1}{2}$ (mod 1). Therefore,

$$h(2^q t) \leq C2^{-\ell_q} . \tag{43}$$

If A is large enough, (43) implies that $h(2^q t) \leq 2^{-\ell_q/2}$ for $\ell_q \geq A$. We can then choose α so that (42) implies that $h(2^q t) \leq 2^{-\alpha\ell_q}$ also when $\ell_q \leq A$. Taking the product, we see that

$$H(t) \leq \prod_{q\in F} h(2^q t) \leq 2^{-\alpha \sum_{q\in F} \ell_q} , \quad \text{and}$$

so (41) will follow at once if we prove that $\sum_{q\in F} \ell_q \geq j$.

The proof is easy : split the interval $\{2,3,\ldots j+1\}$ into intervals I_ℓ which are maximal and such that the sequence α_q is constant on I_ℓ. Each I_ℓ starts at $q+2$ for a $q \in F$ and its length is less than l_q. Also, $j = \sum |I_\ell|$, and so $\sum_F \ell_q \geq j$, as we wanted.

This proves (41), which implies Lemma 7.8, which was our last missing piece for the proof of Theorem 7.1, which is thus established.

8. Properties of the wavelet transform, extensions.

It is probably a good idea to justify all the work done in the previous sections by a few comments on the advantages and applications of the wavelet transforms. However, for reasons of relative conciseness, we do not wish to give anything close to a complete description of all applications of wavelets . Such a description can be found in [My4], or in the references therein. What we'll do instead is a short impressionistic picture of some of the advantages of wavelets .

For convenience, we shall use the notations of dimension 1, and call

$$\psi_{j,k} = 2^{j/2}\psi(2^j x - k) \, , \; j,k \in \mathbf{Z}$$

an orthonormal basis of $L^2(\mathbf{R})$. With these notations, every $f \in L^2(\mathbf{R})$ can be written

$$f = \sum c_{j,k}\psi_{j,k} \quad \text{where} \tag{44}$$

$$c_{j,k} = \int f(x)\, \overline{\psi}_{j,k}(x) dx \, . \tag{45}$$

For practical purposes, it is often easier to use the formula

$$f = \sum_{j\geq 0}\sum_{k\in\mathbf{Z}} c_{j,k}\psi_{j,k} + \sum_{k\in\mathbf{Z}} \alpha_k\varphi(x-k) \, , \tag{46}$$

where the $c_{j,k}$ are given by (45) and

$$\alpha_k = \int f(x)\overline{\varphi(x-k)}dx . \tag{47}$$

(Note that, if ψ and φ are compactly supported, the $c_{j,k}$ and the α_k can all be computed with a local knowledge of f !)

Suppose that the MSA is very regular (for instance, suppose that ψ is Y. Meyer's wavelet of Section 4, Example 2). The first very nice feature of (44) (or (46)) is that the coefficients $c_{j,k}$ are very small, as j becomes large, when f is smooth (we shall give one or two examples later). To see this, one first proves that ψ has almost as many vanishing moments as derivatives, and then one simply integrates by parts in (45). For instance, if $\|f'\|_\infty \leq 1$, then

$$\begin{aligned}
c_{j,k} &= 2^{j/2} \int f(x)\, \overline{\psi(2^j x - k)} dx \\
&= 2^{j/2} \int [f(x) - f(k/2^j)]\, \overline{\psi(2^j x - k)} dx \\
&\leq 2^{-j/2} \int |2^j x - k|\, |\psi(2^j x - k)| dx \\
&\leq C\, 2^{-3j/2} .
\end{aligned}$$

As a general rule, each derivative of f allows one to win a factor of 2^{-j} on the size of $c_{j,k}$.

A slightly more surprising fact is that many of the currently used function spaces can be characterized by the behaviour of the $c_{j,k}$'s (as $j \to +\infty$). This is surprising if one compares the "wavelet transform" to the Fourier transform, for which many such characterizations are false. On the other hand, the reader who has been exposed to some Littlewood-Paley theory will be less surprised. In some sense, one can say that the "wavelet transform", which uses decompositions of frequencies in dyadic blocs, has all the advantages of Littlewood-Paley theory, with the simplicity of the Fourier transform (due to the fact that the coefficients are unique, for instance). Let us give three examples of characterizations of spaces by the size of the $c_{j,k}$'s ; we'll omit the proofs (they are very close to the usual Littlewood-Paley manipulations).

Example 1. A function f is in the Sobolev space $H^s(\mathbf{R})$ if and only if

$$\sum_j \sum_k (1+2^j)^{2s} |c_{j,k}|^2 < +\infty .$$

The condition would be the same in $\mathbf{R}^n$. The condition also holds, by duality, for $s < 0$.

Example 2. The function f is in $L^p(\mathbf{R})$, $1 < p < +\infty$ (or in H^1, the Hardy space, when $p = 1$) if and only if the following square function $g(x)$ is in $L^p(\mathbf{R})$:

$$g(x) = \left\{ \sum_{\substack{j \;,\; k \\ x \in [k2^{-j}, (k+1)2^j]}} \sum 2^{-j} |c_{j,k}|^2 \right\}^{1/2} .$$

Example 3. The function f is in BMO(**R**) if and only if the $c_{j,k}$'s satisfy the following Carleson measure condition : there is a constant $C \geq 0$ such that, for each dyadic interval I,

$$\sum_{\substack{j \quad , \quad k \\ [k2^{-j},(k+1)2^{-j}] \subset I}} \sum |c_{j,k}|^2 \leq C|I| .$$

Of course, these characterizations are also true, with minor modifications, in $\mathbf{R}^n$. Also, many more function spaces have a characterization like this (in particular, the various Besov spaces). The only ones to be excluded are, more or less, the spaces with a L^1 or a L^∞-norm in their definition. We refer to [LM], or [My4], for a list. Also see [FJ3] and [FJW] for more extensive results in the framework of the φ-transform.

An easy corollary of these characterizations is that, for most function spaces E, with the notable exception of BMO, if $f \in E$, then the partial sums $\sum_{j \leq N} c_{j,k}\psi_{j,k}$ converge to f, when $N \to +\infty$, in the space E. So, for instance, if $f \in H^s(\mathbf{R})$ for some $s > 0$, then the convergence of partial sums, which occurs in H^s, is much faster than what would be expected, say, with the Haar system (where we only know the convergence in L^2). It is probably a very nice feature of wavelets : they are some sort of an universal analysing tool in the sense that one does not need to know in advance how regular f will be to obtain, all the same, a rate of convergence of the partial series which is as fast as the regularity of f allows.

Let us conclude this short section by mentioning the existence of a few extensions of various types.

- Periodic wavelets . It is not hard to find an orthonormal basis of $L^2(\mathbf{T}^1)$ (the 1-dimensional torus), say, with the same sort of structure as a wavelet basis. For instance, one can take the "standard" wavelets and periodize them brutally. See [My4], for instance.

- One can define MSA's on $L^2(\mathbf{R}^2)$, say, that are not given by a tensor product, or even are based on a different group structure (like the one associated with a paving of $\mathbf{R}^2$ by hexagons or triangles). In cases that are not too simple, one has to replace the Fourier transform calculations of Sections 2-6 by a suitable use of Gram's orthogonalization process. Still see [My4], for more information.

- One can still go further, and build wavelets on various domains of $\mathbf{R}^n$. Then the group structure disappears (and so does the Fourier transform), but Gram's process still allows one to build interesting bases. For more details, see [JaM1].

- In some cases, the fact that the $\psi_{j,k}$'s form a basis is not quite needed, and the notion of a "frame" is more relevant. Let us not even say what a frame is, and refer directly to [DbGM], or [My4].

- The list above is not exhaustive, but since we don't want Part I to be much longer than the two others, we'll have to leave the wavelets here (we'll use again some of the ideas in part II, however).

PART II

SINGULAR INTEGRAL OPERATORS

—ooOoo—

1. Introduction and generalities

As with Part I, we do not wish to give a complete description of the subject, but to try to explain in detail a small number of techniques that show up in the study of singular integral operators of "Calderón-Zygmund type". In particular, we'll spend some time discussing a new proof of the Tb-theorem by Coifman and Semmes.

Let us start this part with a short review on definitions and basic properties of Calderón-Zygmund operators.

Definition 1.1. A singular integral operator will be, in these notes, a bounded linear operator from the space $\mathcal{D} = C_c^\infty(\mathbb{R}^n)$ of test-functions to its dual $\mathcal{D}'$ (the space of distributions), such that there is a "standard kernel" (the definition is given below) $K(x,y)$ for which

$$(1) \qquad < Tf, g > = \int\int K(x,y) f(y) g(x) dy\, dx$$

whenever $f, g \in \mathcal{D}$ have disjoint supports.

Here, $< Tf, g >$ denotes the effect of the distribution Tf on the function g. Note that we could also have defined T as a bilinear, continuous form on $\mathcal{D} \times \mathcal{D}$. Finally, (1) is the same thing as asking that $Tf(x)$ is defined, for $x \notin supp\, f$, by $Tf(x) = \int K(x,y) f(y) dy$.

Definition 1.2. A "standard kernel" is a function $K(x,y)$, defined on $\mathbb{R}^n \times \mathbb{R}^n \backslash \{(x,y) : x = y\}$, such that, for some constants $0 < \delta < 1$ and $C_0 \geq 0$,

$$(2) \qquad | K(x,y) | \leq C_0 \, | x - y |^{-n}$$

and

$$(3) \qquad | K(x,y) - K(x',y) | + | K(y,x) - K(y,x') | \leq C_0 \frac{| x' - x |^\delta}{| x - y |^{n+\delta}}$$

for $| x' - x | < \frac{1}{2} | x - y |$.

Remarks.

• One could be a little more general than this when defining "standard kernels", but there are not so many examples where that generality would help.

• If T is a singular integral operator (we'll also say a SIO), there is only one kernel associated to T. However, the same kernel can be associated to many different operators. [EXERCISE : show that the operator of multiplication by a given function, or any differential operator are SIO's associated with $K \equiv 0$].

• For the Tb-theorem, a slightly different definition is more natural : given two bounded functions b_1 and b_2, we shall ask that T be defined from $b_1\mathcal{D}$ to $(b_2\mathcal{D})'$ (or be a bilinear form defined on $b_1\mathcal{D} \times b_2\mathcal{D}$), rather than from $\mathcal{D}$ to $\mathcal{D}'$. The rest of the definition is the same.

• Note the nice invariance properties of the class of SIO's by the action of translations and dilations.

Example 1.3. (Principal value operator defined by an antisymmetric kernel). Suppose that $K(x,y)$ satisfies (2), and also $K(x,y) = -K(y,x)$ for all $x \neq y$. Then one can define an operator T (the "principal value operator defined by K") by the formula

$$(4) \qquad < Tf, g >= \frac{1}{2}\int\int K(x,y)[f(y)g(x) - f(x)g(y)]dx\, dy.$$

Exercise : Check that the integral converges because the singularity of K is killed by the fact that $| f(y)g(x) - f(x)g(y) | \leq C(f,g) | x - y |$.

Of course, if K also satisfies (3), then T is a singular integral operator with kernel K. We shall sometimes abbreviate and write $Tf(x) = \text{p. v.} \int K(x,y)f(y)dy$.

Warning : the terminology above is quite convenient for us, but is not used with the same meaning by all mathematicians : in many sources, "singular integrals", "principal values" means something different.

Definition 1.4. (what is meant by T1). Let T be a SIO, and b be a bounded, C^∞ function, but not necessarily compactly supported. Let us define Tb : it will be a distribution known modulo an additive constant, i. e. a continuous linear form defined on the space of functions $g \in \mathcal{D}$ such that $\int g = 0$. Given $g \in \mathcal{D}$ such that $\int g = 0$, write $b = b_1 + b_2$, where $b_1 \in \mathcal{D}$ and b_2 is zero on a neighborhood of $\operatorname{supp} g$. We define $< Tb, g >$ by

$$(5) \qquad \begin{aligned} < Tb, g > &=< Tb_1, g > + \int\int K(x,y)b_2(y)g(x)dy\, dx \\ &=< Tb_1, g > + \int\int \{K(x,y) - K(x_0,y)\}b_2(y)g(x)dy\, dx, \end{aligned}$$

where x_0 is any point in the support of g (the first integral is not actually defined, except in terms of the second one).

Exercise : Check that the second integral in (5) converges because of (3), that the definition does not depend on x_0 or the decomposition $b = b_1 + b_2$, that Tb is a linear function of b, and that it coincides with the definition of Tb when $b \in \mathcal{D}$ or when T is given by an integrable kernel.

Let us now quote, without proofs, a few classical results.

Definition 1.5. If the SIO T can be extended into a bounded operator of $L^2(\mathbb{R}^n)$ (or, in short, if T is bounded on $L^2(\mathbb{R})$), we'll say that T is a Calderón-Zygmund operator (or, more briefly, a "CZO").

Let us say again that this definition is not universally used.

Theorem 1.6. *If T is a CZO, then T has bounded extensions from $L^p(\mathbb{R}^n)$ to $L^p(\mathbb{R}^n)$ for $1 < p < +\infty$, from the Hardy space $H^1(\mathbb{R}^n)$ to $L^1(\mathbb{R}^n)$, and from $L^\infty(\mathbb{R}^n)$ to $BMO(\mathbb{R}^n)$ (defined below).*

This theorem is due to Calderón and Zygmund for the L^p-boundedness, and to Peetre, Spanne and Stein for the L^∞-BMO boundedness.

Definition 1.7. A locally integrable function f is in $BMO(\mathbb{R}^n)$ if $\| f \|_{BMO} = \sup\limits_Q \frac{1}{|Q|} \int_Q |$ $f(x) - f_Q \mid dx < +\infty$, where the supremum is taken over all cubes $Q \subset \mathbb{R}^n$, and $f_Q = \frac{1}{|Q|} \int_Q f(x)dx$.

Remarks

1. A little more is true. Define truncated operators by

$$T_\epsilon f(x) = \int_{|x-y|>\epsilon} K(x,y)f(y)dy \quad \text{for } \epsilon > 0 \text{ and } f \in L^2, \tag{6}$$

and then the maximal operator

$$T_* f(x) = \sup_{\epsilon > 0} | T_\epsilon f(x) | . \tag{7}$$

If T is a CZO, then T_* is bounded on L^p, $1 < p < +\infty$, and in particular the T_ϵ's are uniformly bounded. This result is obtained by means of "Cotlar's inequality" (we shall actually prove this inequality, in a more general context, in Part III).

2. If the T_ϵ's are uniformly bounded on L^2, say, then one can extract a sequence that converges weakly to some bounded operator $\tilde{T}$. It could be that T is not bounded (for instance, if T is a differential operator). If T is bounded, then $T - \tilde{T}$ is bounded, too, has kernel 0, and so (exercise !) $T - \tilde{T}$ is the operator of pointwise multiplication by a bounded function [Hint for the exercise : show that the function is, locally, $(T - \tilde{T})(\mathbb{1}_{[-N,N]^n})$].

Note that the T_ϵ's do not necessarily converge when $\epsilon \to 0$: consider the one-dimensional kernel $K(x,y) = | x - y |^{-1+i\gamma}$.

3. If T is a CZO, and $T1 = 0$ (modulo constants), then Theorem 1.6 can be slightly improved : T is also bounded from BMO to BMO. Also, if T is a CZO such that $T^t 1 = 0$, then T is bounded on H^1 (the transposed operator T^t is defined by $< T^t f, g > = < Tg, f >$; it is of course a SIO with kernel $K(y, x)$).

We have seen already that our lack of knowledge on the restriction of the "distribution-kernel" of T to the diagonal is a potential source of trouble. In particular, it authorizes behaviors that do not have the correct invariance under the action of translations and dilations (think about the pointwise multiplication by a wild function, or $Tf = f'$). The following notion of "weak boundedness" was introduced by Y. Meyer to try and compensate this lack of knowledge.

Definition 1.8. For $t > 0$ and $x \in \mathbb{R}^n$, call $\mathcal{A}_{t,x}$ the operator of translation by x and dilation by t defined by $\mathcal{A}_{t,x} f(u) = t^{n/2} f(tu - x)$.

We shall say that the operator $T : \mathcal{D} \to \mathcal{D}'$ is weakly bounded if the operators $\mathcal{A}_{t,x}^{-1} T \mathcal{A}_{t,x}$ are uniformly bounded from $\mathcal{D}$ to $\mathcal{D}'$ for $t > 0$ and $x \in \mathbb{R}^n$. This means that, for each compact set K, there is a suitable seminorm $|||\cdot|||$ such that

$$|< \mathcal{A}_{t,x}^{-1} T \mathcal{A}_{t,x} f, g >| \leq C \, ||| f ||| \, ||| g |||$$

for all f and g supported in K.

Let us give an equivalent definition. For each $B \subset \mathbb{R}^n$ and each integer $q \geq 0$, define, for f supported in B,

$$N_q^B(f) = R^{n/2} \sum_{|\alpha| \leq q} R^{|\alpha|} \, \| \partial^\alpha f \|_\infty, \tag{8}$$

where R is the radius of B. Note that if $B = B(x_0, R)$ and f is supported in B, then $N_q^B f = N_q^{1/2B} g$, where $g(x) = 2^{n/2} f(2x - x_0)$. We can use this invariance, and invariance by translation, to assert that T is weakly bounded if and only if there exist $q \geq 0$ and $C \geq 0$ such that, for all balls B and all functions $f, g \in \mathcal{D}$ supported in B,

$$|< Tf, g >| \leq C N_q^B(f) N_q^B(g). \tag{9}$$

Exercises

1. Check what was just said.

2. Show that if T is bounded on L^2 (or L^p), then T is weakly bounded.

3. Check that the operator defined by $Tf = f'$ in dimension 1 is not weakly bounded.

4. Show that $Tf = mf$ (for some $m \in L^1_{\text{loc}}(\mathbb{R}^n)$) is weakly bounded if and only if $m \in L^\infty$.

5. Show that if the kernel $K(x,y)$ satisfies (2) and is antisymmetric (i.e., $K(y,x) = -K(x,y)$), then the principal value operator defined by (4) is weakly bounded (we shall use this exercise later).

Remark. If T is associated by (1) to a kernel $K(x,y)$ verifying (2), and if (9) is satisfied for some $q > 0$, one can show that it is true for all $q > 0$ (with suitable modifications when q is not an integer). The proof is not too hard, but takes some space. This means that T is also defined as a bilinear form on any space $C^\epsilon = \{f \in L^\infty : |f(x) - f(y)| \leq C|x-y|^\epsilon$ for all $x, y\}$. We could have changed the definitions above to take this into account (this is actually what happens on spaces of homogeneous type where C^∞ does not make sense), but we'll often find it more convenient to define T on test-functions.

We now have enough definitions to state the "T1-theorem".

Theorem 1.9 ([DJé]). *Let T be a singular integral operator (like in Definition 1.1). Then T has a bounded extension to $L^2(\mathbb{R}^n)$ if and only if*

i) T *is weakly bounded (see Definition 1.8) ;*
ii) $T1 \in BMO$ *(see Definition 1.4) ;*
iii) $T^t 1 \in BMO$.

Recall that T^t is the SIO defined by $< T^t f, g > = < Tg, f >$. The direct implication follows from Peetre, Spanne and Stein's result (Theorem 1.6) and the exercise 2 above. We shall prove this theorem later, and we'll discuss applications later still. Before, let us state a second result (the "Tb-theorem").

Definition 1.10 (Paraaccretivity). Let b be a bounded function from $\mathbb{R}^n$ to $\mathbb{C}$.

We'll say that b is accretive if there is a $\delta > 0$ such that $\mathcal{R}e\, b(x) \geq \delta$ a. e.

We'll say that b is "paraaccretive" if there are constants $C \geq 0$ and $\delta > 0$ such that, for all $x \in \mathbb{R}^n$ and $r > 0$, there exists a cube Q such that $\operatorname{dist}(x, Q) \leq Cr$, $\frac{1}{C} r \leq \operatorname{diam} Q \leq Cr$, and such that

$$\left| \frac{1}{|Q|} \int_Q b(y)\, dy \right| \geq \delta. \tag{10}$$

We'll even say that b is "special paraaccretive" if (10) is true for all dyadic cubes.

Exercise. Check that

- if b is paraaccretive, then $|b| \geq \delta$ a. e. ;
- in dimension 1, e^{ix} is not paraaccretive ;
- cubes can be replaced by balls in the above definition, without changing the notion.

The next theorem says that, in Theorem 1.9, the function 1 can be replaced by any paraaccretive function b. The definitions are slightly more complicated, because it is more natural to define T from $b\mathcal{D}$ to $(b\mathcal{D})'$ in this case.

Notation. If $b \in L^\infty$, we shall denote by M_b the operator of pointwise multiplication by b.

Theorem 1.11 ([DJS]). *Let b_1 and b_2 be two paraaccretive functions. Let T be a (bounded) operator from $b_1\mathcal{D}$ to $(b_2\mathcal{D})'$. Suppose that T is associated to a standard kernel $K(x,y)$ in the sense that $< Tf, g >= \int\int K(x,y)f(y)g(x)dy\,dx$ whenever $f \in b_1\mathcal{D}$ and $g \in b_2\mathcal{D}$ have disjoint supports.*

Also suppose that $M_{b_2}TM_{b_1}$ is weakly bounded, that $Tb_1 \in BMO$ and $T^tb_2 \in BMO$. Then T extends to a bounded operator on $L^2(\mathbb{R}^n)$.

This theorem is actually a generalization of a result by A. McIntosh and Y. Meyer [McM] : the important special case when $Tb_1 = T^tb_2 = 0$ for accretive functions b_1 and b_2. Their proof uses a formulation in terms of interpolation of the L^2-boundedness of the Cauchy integral on Lipschitz graphs [CMM].

Remarks

1. We didn't define Tb_1 (or T^tb_2) in this case. This time, it will be a linear form on $b_2\mathcal{D}$, defined modulo an additive constant. This means that $< Tb_1, b_2\varphi >$ will be defined if φ is a test function such that $\int \varphi(x)b_2(x)dx = 0$. The definition of $< Tb_1, b_2\varphi >$ is similar to what was done in Definition 1.4. One picks a function $\mathcal{X} \in \mathcal{D}$ such that $\mathcal{X} \equiv 1$ on a neighborhood of $\operatorname{supp}\varphi$, and one writes

$$< Tb_1, b_2\varphi >=$$

$$< T(b_1\mathcal{X}), b_2\varphi > + \int\int K(x,y)[b_1(1-\mathcal{X})](y)b_2(x)\varphi(x)dy\,dx$$

$$=< T(b_1\mathcal{X}), b_2\varphi > + \int\int \{K(x,y) - K(x_0,y)\}\{b_1(1-\mathcal{X})\}(y)b_2(x)\varphi(x)dy\,dx.$$

The last integral converges because of (3), as before.

2. For both theorems, one actually gets a control on T, like $\| T \|_{L^2,L^2} \leq k \| Tb_1 \|_{BMO} + k \| T^tb_2 \|_{BMO} + kC_0 + kC$, where C_0 is the constant of (2) and (3), and C comes from the weak boundedness (C is any constant that shows up in an inequality (9)). This easily follows from the proof, or from Baire's Category theorem.

3. The converse of Theorem 1.11 is, as for the T1-theorem, a direct consequence of Theorem 1.6.

4. Paraaccretivity is, in some sense, an optimal notion : one can show that if b is such that the theorem is true for $b_1 = b_2 = b$ (and all operators !), then b is paraaccretive (see [DJS]).

The next sections are devoted to giving a short, new proof of the Tb-theorem which is due to R. Coifman and S. Semmes [CJS]. It is also a nice coincidence that it uses ideas related to wavelets.

2. First step of Coifman-Semmes' proof : a Riesz basis

This section is needed only for the Tb-theorem. If $b_1 = b_2 = 1$, then we shall construct the Haar system in this section !

Given a paraaccretive function b, we intend to build a Riesz basis of $L^2(\mathbb{R}^n)$. We shall first consider the case of a "special paraaccretive" function b, i.e. a bounded function such that, for some $\delta > 0$, (10) is true for all dyadic cubes.

Let us first define projection operators : for $k \in \mathbb{Z}$, let E_k be such that

$$E_k f(x) = \frac{1}{|Q|}\int_Q f(t)dt, \tag{11}$$

where Q is the dyadic cube of side length 2^{-k} that contains x.

(This is the projection operator associated to the Haar system ; note that it is a typical martingale projection.) Also define the difference $D_k = E_{k+1} - E_k$. Now define corresponding projection operators "relative to b" :

$$F_k f(x) = \frac{1}{E_k b(x)} E_k(bf)(x) = \{\int_Q b(t)dt\}^{-1}\int_Q f(t)b(t)dt, \tag{12}$$

where again Q is the dyadic cube of size 2^{-k} containing x ; note that the first denominator is bounded below because of (10).

Finally, call $\Delta_k = F_{k+1} - F_k$.

Let us check a few easy facts.

First, note that $\int_Q f(t)b(t)dt = \int_Q F_k f(t)b(t)dt$ for all dyadic cubes Q of size 2^{-k} (the "martingale property"). Next,

$$F_j F_k = F_{j\wedge k}. \tag{13}$$

If $k \leq j$, then $F_k f$ is constant on each cube of size 2^{-k}, and so it is constant on cubes of size 2^{-j}, and F_j does not change it. If $k > j$, we use the martingale property above for F_k : the integral of F_k on cubes of size 2^{-k} is the same as the integral of f ; this remains true for cubes of size 2^{-j} and so $F_j F_k f = F_j f$.

$$\Delta_j \Delta_k = \delta_{j,k}\Delta_j. \tag{14}$$

Just expanding gives $\Delta_j\Delta_k = (F_{j+1} - F_j)(F_{k+1} - F_k) = F_{j+1}F_{k+1} - F_jF_{k+1} - F_{j+1}F_k + F_jF_k$.

If $j > k$, what remains is $F_{k+1} - F_{k+1} - F_k + F_k = 0$; if $j < k$, one also gets 0 (by symmetry) and if $j = k$, one is left with $F_{j+1} - F_j - F_j + F_j = \Delta_j$.

If u and v are compactly supported and $k \neq \ell$,

$$\int (\Delta_k u)(\Delta_\ell v)b\,dx = 0. \tag{15}$$

We can assume $k > \ell$. Then $\Delta_\ell v$ is constant on each cube of size 2^{-k}. On such a cube Q, F_{k+1} and F_k preserve the integral against b, and so $\int_Q (\Delta_k u) b = 0$. We just have to sum on Q.

Let us now prove a square function estimate.

Lemma 2.1. *There is a constant $C \geq 0$ such that, for all $f \in L^2$,*

$$C^{-1} \| f \|_2^2 \leq \int_{\mathbb{R}^n} \sum_{k \in \mathbb{Z}} | \Delta_k f(x) |^2 \, dx \leq C \| f \|_2^2 . \tag{16}$$

Let us first prove the second inequality. Write

$$\begin{aligned} \Delta_k f &= (E_{k+1} b)^{-1} E_{k+1}(bf) - (E_k b)^{-1} E_k(bf) \\ &= [(E_{k+1} b)^{-1} - (E_k b)^{-1}] E_{k+1}(bf) + (E_k b)^{-1} [E_{k+1}(bf) - E_k(bF)]. \end{aligned}$$

Since $[E_k b \, E_{k+1} b]^{-1}$ is bounded because of (10), we get

$$| \Delta_k f |^2 \leq C \, | D_k b |^2 | E_{k+1}(bf) |^2 + C \, | D_k(bf) |^2$$

and

$$\begin{aligned} \int \sum | \Delta_k f |^2 &\leq C \int \sum | D_k b |^2 | E_{k+1}(bf) |^2 + C \int \sum | D_k(bf) |^2 \\ &\leq I + II. \end{aligned}$$

For the second part, we use the usual square function estimate (even simplified) : note that $bf = \sum_k D_k(bf)$, and that the pieces are orthonormal to each other because D_k is self-adjoint and $D_k D_\ell = \delta_{k,\ell} D_k$. So $\| bf \|_2^2 = \sum_k \| D_k(bf) \|_2^2 = \int \sum | D_k(bf) |^2$.

For the first part, we use the fact that $| D_k b |^2 \, dx$, times the counting measure on $\{2^{-k} : k \in \mathbb{Z}\}$, is a Carleson measure. To check this easy fact, one must prove that, if Q is a cube and R is its sidelength,

$$\int_Q \sum_{k : 2^{-k} \leq R} | D_k b(x) |^2 \, dx \leq C \, | Q | .$$

But the left hand side is equal to

$$\begin{aligned} \int_Q \sum_{2^{-k} \leq R} | D_k \{ b(t) \mathbb{1}_{CQ}(t) \}^2 dx &\leq \int_Q \sum_k | D_k (b \mathbb{1}_{CQ}) |^2 \, dx \\ &\leq C \, \| b \mathbb{1}_{CQ} \|_2^2 \quad \text{(by the standard square function estimate)} \\ &\leq C \, \| b \|_\infty | Q |, \text{ as needed.} \end{aligned}$$

The fact that, if $\sum_k | D_k b |^2 \, dx\, d\delta_{2^{-k}}$ is a Carleson measure, then

$$\int \sum | D_k b |^2 | E_{k+1}(bf) |^2 \, dx \leq C \| bf \|_2^2 \leq C \| f \|_2^2$$

is a standard fact about Carleson measures (see Th. I.5.6 in [Ga3]).

So we proved the second inequality. The first one will follow by duality. Note that b^{-1} is bounded because b is paraaccretive, and

$$\begin{aligned} \| f \|_2^2 &= \int (b^{-1} f) f b = \sum_k \sum_\ell \int \Delta_\ell (f b^{-1}) \Delta_k (f) b \, dx \\ &= \sum_k \int \Delta_k (f b^{-1}) \Delta_k (f) b \, dx \quad \text{(by (15))}. \end{aligned}$$

If you are worried about the convergence, note that the formula is true when f is a finite linear combination of characteristic functions of dyadic cubes ; we'll be able to pass to the limit easily, since both sides of the inequality are continuous in L^2–norm. Apply Schwarz' inequality :

$$\begin{aligned} \| f \|_2^2 &\leq C \left\{ \int \sum_k | \Delta_k (f b^{-1}) |^2 \right\}^{1/2} \left\{ \int \sum_k | \Delta_k f |^2 \right\}^{1/2} \\ &\leq C \| f \|_2 \left\{ \int \sum_k | \Delta_k f |^2 \right\}^{1/2}. \end{aligned}$$

Dividing by $\| f \|_2$ if it is not zero, one gets the first inequality of Lemma 2.1. So this lemma is true.

Remember that we want to find a Riesz basis that looks like the Haar system, but would be "orthogonal" with respect to the b–scalar product. One should be careful : if b is not real, this is just an analogy, and I heard from a few short proofs that were a little too short precisely because the analogy was pushed a little too far.

For each cube Q of the kth generation (i.e., each dyadic cube of side length 2^{-k}), call Q_η, $\eta \in I$, the cubes of the $(k+1)^{st}$ generation that are contained in Q (here, I has 2^n elements). For each η, call $\beta_\eta = \int_{Q_\eta} b(x) dx$. By (10), $| \beta_\eta | \geq 2^{-n} \delta | Q |$.

Let us call 0 one of the elements of I, and $\hat{I} = I \backslash \{0\}$. We want to define, for each $\epsilon \in \hat{I}$, a function h_Q^ϵ which is constant on each Q_η, supported on Q, and has the properties

$$\int_Q h_Q^\epsilon(x) b(x) dx = 0, \tag{17}$$

$$\int_Q h_Q^\epsilon(x) h_Q^{\epsilon'}(x) b(x) dx = 0 \quad \text{for all } \epsilon' \neq \epsilon \text{ in } \hat{I}, \tag{18}$$

and the normalization

(19) $$\int_Q \left[h_Q^\epsilon(x)\right]^2 b(x)dx = 1.$$

Exactly like in Part I, Section 6, it is more convenient to formulate the question in terms of matrices. For $\epsilon \in \hat{I}$ and $\eta \in I$, call $\alpha_{\epsilon,\eta}$ the value of h_Q^ϵ on Q_η. Let us complete the matrix by taking $\alpha_{0,\eta} = \left(\int_{Q_\eta} b(x)dx\right)^{-1/2}$ for all η. Also call v_ϵ, $\epsilon \in I$, the vector with coordinates $(\alpha_{\epsilon,\eta})_{\eta\in I}$. The conditions (17) and (18) say that the v_ϵ, $\epsilon \in I$, are mutually "orthogonal" for the bilinear form $< (a_\eta),(b_\eta) >_\beta = \sum_\eta a_\eta b_\eta \beta_\eta$. We are given a first vector v_0, and have to "complete the basis".

First consider $v_0^\perp = \{w = (w_\eta) \ : \ \sum w_\eta \alpha_{0,\eta}\beta_\eta = 0\}$. This is a space of dimension $2^n - 1$; the restriction to $v_0^\perp$ of the bilinear form $< , >_\beta$ cannot be zero, because it would imply that each element of $v_0^\perp$ is "orthogonal" to the whole space (but the linear form $< w, \cdot >_\beta$ does not identically vanish unless $w = 0$). So one can find two elements w_1 and w_2 of $v_0^\perp$, such that $< w_1, w_2 >_\beta \neq 0$. If both $< w_1, w_1 >_\beta$ and $< w_2, w_2 >_\beta$ vanish, then $< w_1 + w_2, w_1 + w_2 >_\beta \neq 0$. In any case, we can find a vector $v_1 \in v_0^\perp$ such that $< v_1, v_1 >_\beta \neq 0$ and, after normalization, one can even assume that $< v_1, v_1 >_\beta = 1$ (we are implicitly assuming here that $I = \{0, 1, \cdots\}$; otherwise, we would give v_1 is a more complicated name).

Let us do the same thing again : $v_0^\perp \cap v_1^\perp$ is a $(2^n - 2)$-dimensional space, and (unless $n = 1$) the bilinear form $< , >_\beta$ cannot vanish on this whole space. So we can find w_1 and $w_2 \in v_0^\perp \cap v_1^\perp$ such that $< w_1, w_2 >_\beta \neq 0$. Then there is a $v_2 \in v_0^\perp \cap v_1^\perp$, with $< v_2, v_2 >_\beta = 1$. Applying this argument enough times, one gets vectors v_ϵ, $\epsilon \in I$, such that $< v_\epsilon, v_{\epsilon'} >_\beta = \delta_{\epsilon,\epsilon'}$. Translating things back in terms in terms of functions, we just proved the existence of the h_Q^ϵ's.

Note that the v_ϵ's are independent : if $\sum \lambda_\epsilon v_\epsilon = 0$, then taking the twisted scalar product with $v_{\epsilon'}$ gives $\lambda_{\epsilon'} = 0$. Hence, the h_Q^ϵ's are independent from each other, and are also independent from the constant function (coming from v_0).

Lemma 2.2. *Call V_Q the space of all functions f that are supported on Q, constant on each Q_η, and such that $\int fb = 0$. Then every $f \in V_Q$ can be written*

(20) $$f = \sum_{\epsilon\in\hat{I}} < f, h_Q^\epsilon >_b h_Q^\epsilon,$$

where $< f, g >_b = \int fgb$ is a notation for the twisted scalar product. Also, if the h_Q^ϵ's are correctly chosen,

(21) $$C^{-1} \parallel f \parallel_2^2 \leq \sum_{\epsilon\in\hat{I}} |< f, h_Q^\epsilon >_b|^2 \leq C \parallel f \parallel_2^2$$

for all $f \in V_Q$, and a constant C that does not depend on Q or ϵ.

To prove (20), just note that, since the h_Q^ϵ's are independent, they form a basis of V_Q. So every $f \in V_Q$ can be written $f = \sum C_\epsilon h_Q^\epsilon$, and then

$$\int f h_Q^{\epsilon'} b = \sum_\epsilon C_\epsilon \int h_Q^\epsilon h_Q^{\epsilon'} b = C_{\epsilon'}$$

(by (18) and (19)), which gives (20).

A short glance at the construction of the h_Q^ϵ's is enough to convince oneself that one can choose the v_ϵ's so that their coordinates (the $\alpha_{\epsilon,\eta}$'s) are less than

$$C \left\{ \min_\eta \mid \beta_\eta \mid \right\}^{-1/2} \leq C \mid Q \mid^{-1/2}.$$

Consequently, one can choose that h_Q^ϵ's so that $\| h_Q^\epsilon \|_\infty \leq C \mid Q \mid^{-1/2}$. The second inequality of (21) is then trivial, and the first one follows from (20). This proves Lemma 2.2.

Lemma 2.3. *The h_Q^ϵ's, where Q is a dyadic cube and $\epsilon \in \hat{I}$, are a Riesz basis of $L^2(\mathbb{R}^n)$. More precisely, each function $f \in L^2$ can be written*

(22) $$f = \sum_Q \sum_\epsilon < f, h_Q^\epsilon >_b h_Q^\epsilon,$$

and

(23) $$C^{-1} \| f \|_2^2 \leq \sum_Q \sum_\epsilon |< f, h_Q^\epsilon >_b|^2 \leq C \| f \|_2^2 .$$

By a simple density argument, we only have to prove (22) and (23) for a dense set, like the finite linear combinations of characteristic functions of dyadic cubes. For such a function f, $f = \sum_k \Delta_k f$, with $\| f \|_2^2 \sim \sum_k \| \Delta_k f \|_2^2$ by Lemma 2.1. Let us now fix k, and decompose $g_k = \Delta_k f$ a little further. By definition of Δ_k, g_k is constant on each cube of generation $k+1$, and its integral on each cube of generation k is zero. So we can write $g_k = \sum_Q \mathbb{1}_Q g_k$, where one sums on all dyadic cubes of sidelength 2^{-k}, and then $\mathbb{1}_Q g_k \in V_Q$.

So $g_k = \sum_Q \left\{ \sum_\epsilon < \mathbb{1}_Q g_k, h_Q^\epsilon >_b h_Q^\epsilon \right\}$ by Lemma 2.2. But $< \mathbb{1}_Q g_k, h_Q^\epsilon >_b = < g_k, h_Q^\epsilon >_b =$ $< F_{k+1} f, h_Q^\epsilon >_b - < F_k f, h_Q^\epsilon >_b = < F_{k+1} f, h_Q^\epsilon >_b$ (since $F_k f$ is constant on Q and h_Q^ϵ satisfies (17)). This is equal to $< f, h_Q^\epsilon >_b$ (because h_Q^ϵ is constant on each Q_η). Hence (22) is true.

Furthermore, $\| g_k \|_2^2 = \sum_{Q \text{ of generation } k} \| \mathbb{1}_Q g_k \|_2^2 \sim \sum_Q \sum_\epsilon |< \mathbb{1}_Q g_k, h_Q^\epsilon >|^2$ (by (21)) $= \sum_{Q \text{ of generation } k} \sum_\epsilon |< f, h_Q^\epsilon >|^2$, from which (23) follows.

Up to now, we associated to each special-paraaccretive function b a Riesz basis (h_Q^ϵ), with the following nice properties : for each Q and each ϵ, h_Q^ϵ is supported on Q, and is constant on each cube of the next generation ; also, the coefficient of f relative to h_Q^ϵ is the twisted scalar product $\int f(x)h_Q^\epsilon(x)b(x)dx$.

To conclude this section, let us indicate how the construction above can be modified in the general case (when we only know that b is paraaccretive). The idea will be to keep the same formalism, and modify the dyadic cubes. The following lemma will help us take care of the geometry.

Lemma 2.4. *Let b be a paraaccretive function. There exist constants $A \in \mathbb{N}$ and $\delta' > 0$ such that, if $x \in \mathbb{R}^n$ and $k \in \mathbb{Z}$, there is a dyadic cube Q, of sidelength 2^k, at distance $\leq 2^{A+k}$ from x, and such that*

$$\left|\frac{1}{|Q|}\int_Q b\right| \geq \delta'. \tag{24}$$

This is a very easy consequence of Definition 1.10, and is left as an exercise.

Let $\mathcal{R}$ denote the set of 2^{A+2}–adic cubes, and $\mathcal{R}_0 \subset \mathcal{R}$ the subset composed of the "good" cubes, i.e. cubes Q_0 such that $\left|\frac{1}{|Q_0|}\int_{Q_0} b\right| \geq \frac{1}{2}\delta' 2^{-(A+2)n}$, where δ' is as in Lemma 2.4. If $Q_0 \in \mathcal{R}\backslash\mathcal{R}_0$ is not good, we can use Lemma 2.4 to find a cube $Q \in \mathcal{R}$ contained in Q_0, with $\text{length}(Q) = 2^{-A-2}\text{length}(Q_0)$, and satisfying (24). Call $A(Q_0) = Q$, and $B(Q_0) = Q_0\backslash Q$; since Q_0 was not a good cube, we get

$$\left|\frac{1}{B(Q_0)}\int_{B(Q_0)} b\right| \geq \frac{1}{2}\delta' 2^{-(A+2)n}.$$

Our new notion of "cubes" is defined as follows : we shall call "cubes of generation k" all cubes of $Q_0 \in \mathcal{R}_0$ with sidelength $2^{-k(A+2)}$, and all sets of the form $A(Q_0)$ or $B(Q_0)$ for a Q_0 of sidelength $2^{-k(A+2)}$ which is in $\mathcal{R}\backslash\mathcal{R}_0$. Note that, if $\delta'' = \frac{1}{2}\delta' 2^{-(A+2)n}$, then every "cube of generation k" R satisfies

$$\left|\frac{1}{|R|}\int_R b\right| \geq \delta''. \tag{25}$$

Also, all the cubes of a given generation have roughly the same size, and they form a partition of $\mathbb{R}^n$. Finally, if R is a cube of generation k, all the cubes of generation $k+1$ that meet R are contained in R, and the number of these cubes is $\leq C$.

We can now apply the same argument as in the beginning of this section, and get a Riesz basis h_Q^ϵ, where this time Q runs in the new set of cubes, and for each Q, ϵ is in a set $\hat{I}(Q)$ (which this time does not always have the same number of elements). The formulas (22) and (23), in particular, are still true in this context.

3. Proof of Theorem 1.11 when $Tb_1 = T^t b_2 = 0$

To simplify notations a little, we shall do the proof when $b_1 = b_2 = b$ (we'll say a few words later on how to modify the argument in the general case). We'll also suppose that b is "special paraaccretive", the modifications to obtain the general case would be trivial.

The idea of the proof is, simply, to estimate the coefficients of the matrix of TM_b in the basis (h_Q^ϵ) of the previous section, and to show that they decrease rapidly enough away from the diagonal. Let us isolate in a first lemma the only part of the proof where we'll need the regularity estimate (3) on the kernel of T.

Lemma 3.1. *Let Q be a dyadic cube, and let x_Q be its center. Let h be supported on Q, and such that $\int_Q h\, b = 0$. Then, for $x \notin 2Q$,*

$$(26) \qquad |\tilde{T}h(x)| \leq C\, |Q|^{1+\frac{\delta}{n}} |x - x_Q|^{-n-\delta} \| h \|_\infty,$$

where $\tilde{T} = M_b T M_b$. In particular, if h is one of the h_Q^ϵ's, we get

$$(27) \qquad |\tilde{T}h_Q^\epsilon(x)| \leq C\, |Q|^{\frac{1}{2}+\frac{\delta}{n}} |x - x_Q|^{-n-\delta} \quad \textit{for } x \notin 2Q.$$

To prove the lemma, write

$$\tilde{T}h(x) = b(x) \int_Q K(x,y) b(y) h(y)\, dy$$
$$= b(x) \int_Q [K(x,y) - K(x,x_Q)] b(y) h(y)\, dy,$$

and (26) follows from (3) ; (27) follows from (26) and the fact that $\| h_Q^\epsilon \|_\infty \leq C\, |Q|^{-1/2}$.

Remark 3.2. This is the only place where (3) is used (we'll use this remark in Section 4).

Call $C_{Q,R}^{\epsilon,\epsilon'} = < \tilde{T}h_Q^\epsilon, h_R^{\epsilon'} >$. Because h_Q^ϵ and $h_R^{\epsilon'}$ are not smooth, we did not say yet how to define this number. When the closures of Q and R are disjoint, we can of course use the kernel of T ; when T is the principal value operator associated with an antisymmetric kernel, it is easy to check that the formula (4) can still be used. For the general case, we shall have to use a small limiting argument and the weak boundedness of $\tilde{T}$. We postpone this definition to the proof of Lemma 3.4. Since the indices ϵ and ϵ' do not play any role in the estimates, we shall simplify and write $C_{Q,R} = < \tilde{T}h_Q, h_R >$ instead (the ϵ's will reappear only when we need them).

Let us summarize the estimates on the $C_{Q,R}$ that we want.

Lemma 3.3. *Let Q and R be dyadic cubes, and let the R_η's be the cubes of the next generation that are contained in R. Then*

$$(28) \quad \textit{if } |Q| \leq |R|, \textit{ then } |C_{Q,R}| \leq C\, |Q|^{1/2} |R|^{-1/2} \frac{|Q|^{\delta/n}}{|Q|^{\delta/n} + dist(Q, \bigcup_\eta \partial R_\eta)^\delta};$$

(29) *if* $|Q| \leq |R|$ *and* $dist(Q,R) \geq |R|^{1/n}$, *then*

$$|C_{Q,R}| \leq C|Q|^{1/2}|R|^{1/2}\frac{|Q|^{\delta/n}}{dist(Q,R)^{n+\delta}};$$

(30) *if* $|Q| \geq |R|$, *then* $|C_{Q,R}| \leq C|R|^{1/2}|Q|^{-1/2}\dfrac{|R|^{\delta/n}}{|R|^{\delta/n}+dist(R,\bigcup_\eta \partial Q_\eta)^\delta}$;

(31) *if* $|Q| \geq |R|$ *and* $dist(Q,R) \geq |Q|^{1/n}$, *then*

$$|C_{Q,R}| \leq C|Q|^{1/2}|R|^{1/2}\frac{|R|^{\delta/n}}{\mathrm{dist}(Q,R)^{n+\delta}}.$$

It will be clear from the proof of (28) and (29) that (30) and (31) follow, just by exchanging Q and R.

Let us start with the case when $|Q| \leq |R|$, Q is not contained in R, and even $\mathrm{dist}(Q,R) \geq |Q|^{1/n}$. Integrating (27) gives

$$\begin{aligned}|C_{Q,R}| &\leq \int_R C|Q|^{\frac{1}{2}+\frac{\delta}{n}}|x-x_Q|^{-n-\delta}|h_R(x)|\,dx\\ &\leq C|Q|^{\frac{1}{2}+\frac{\delta}{n}}|R|^{-1/2}\int_R |x-x_Q|^{-n-\delta}\,dx\\ &\leq C|Q|^{\frac{1}{2}+\frac{\delta}{n}}|R|^{-1/2}\,\mathrm{dist}(Q,R)^{-\delta},\end{aligned}$$

as wanted for (28). If we even know that $\mathrm{dist}(Q,R) \geq |R|^{\frac{1}{n}}$, then we simply can replace the last line by
$C|Q|^{\frac{1}{2}+\frac{\delta}{n}}|R|^{-1/2}|R|\,\mathrm{dist}(Q,R)^{n+\delta}$, which gives (29).

Next, suppose that $|Q| \leq |R|$, Q is not contained in R, but $\mathrm{dist}(Q,R) \leq |Q|^{\frac{1}{n}}$. In this case, we write $h_R = h_1 + h_2$, where h_1 is supported in $3Q$ and h_2 lives outside $2Q$. The estimate above still gives $|<\tilde{T}h_Q, h_2>| \leq C|Q|^{1/2}|R|^{-1/2}$.

For $<\tilde{T}h_Q, h_1>$, one can use a small limiting argument, which is very similar to the argument in Lemma 3.4 below (but simpler), and thus left as an exercise, to show that

$$\begin{aligned}|<\tilde{T}h_Q,h_1>| &= \left|\int\int K(x,y)h_Q(y)h_1(x)b(x)b(y)\,dx\,dy\right|\\ &\leq C\int\int |x-y|^{-n}|Q|^{-1/2}|R|^{-1/2}\,1_Q(y)1_{3Q\setminus Q}(x)\,dx\,dy\\ &\leq C|Q|^{1/2}|R|^{-1/2} \leq C|Q|^{1/2}|R|^{-1/2}\frac{|Q|^{\delta/n}}{|Q|^{\delta/n}+\mathrm{dist}(Q,\bigcup_\eta R_\eta)^\delta}\end{aligned}$$

and complete the proof of (28) in this second case.

We are now left with the case when Q is contained in R. Let us start with the case when Q is strictly contained in R, and call R_0 the R_η which contains Q. Note that h_R

is constant on R_η. We can use the fact that $\tilde{T}^t 1 = 0$ (or, equivalently, $T^t b = 0$), so that $< \tilde{T} h_Q, h_R > = < \tilde{T} h_Q, h_R - c >$, where c is the value of h_R on R_0. Note that this is not quite conform to our definition of $T^t b$ (the first remark after the statement of Theorem 1.11) : we said that $< T(bh_Q), b >$ is defined when h_Q is a test function such that $\int h_Q b = 0$. Here, we still have $\int h_Q b = 0$, but h_Q is not smooth. A small covering argument allows one to extend the definition to the case when h_Q is a finite sum of characteristic functions of cubes (the idea is that in this case, the singularity of h_Q is not too bad). Once again, we leave the details to the reader, because we shall give a slightly more complete argument in Lemma 3.4 below.

Note that $h_R - c$ vanishes on $R_0 \supset Q$, and so we'll be able to use the kernel again :

$$\begin{aligned} | C_{Q,R} | &\leq C \int_{2Q \cap R_0^c} | \tilde{T} h_Q(x) || R |^{-1/2} \, dx + \int_{(2Q)^c \cap R_0^c} | \tilde{T} h_Q(x) || R |^{-1/2} \, dx \\ &\leq C \int\int | Q |^{-1/2} | R |^{-1/2} | x - y |^{-n} \, 1_Q(y) 1_{2Q \cap R_0^c}(x) \, dy \, dx \\ &\quad + C | Q |^{\frac{1}{2} + \frac{\delta}{n}} | R |^{-1/2} \int_{(2Q)^c \cap R_0^c} | x - x_Q |^{-n-\delta} \, dx \end{aligned}$$

(we used (2) and (27)).

Therefore,

$$| C_{Q,R} | \leq C | Q |^{1/2} | R |^{-1/2} + C | Q |^{\frac{1}{2} + \frac{\delta}{n}} | R |^{-1/2} \{ \mathrm{dist}(Q, R_0^c) + | Q |^{\frac{1}{n}} \}^{-\delta},$$

and so (28) is also true in this case (the first term disappears when $2Q \cap R_0^c$ is empty).

The inequality (28) will be completely established as soon as we prove it when $Q = R$. Since (30) and (31) are a trivial corollary of the proof, Lemma 3.3 will be a direct consequence of the following lemma.

Lemma 3.4. $| < \tilde{T} h_Q, h_Q > | \leq C$.

Note that h_Q can be written $h_Q = \sum \lambda_i 1_{R_i}$, where the sum has less than C terms, the R_i's are disjoint cubes of roughly the same size as Q, and $| \lambda_i | \leq C | Q |^{-1/2}$. Since $| < \tilde{T} 1_{R_i}, 1_{R_j} > | \leq C | Q |$ when $i \neq j$ by a brutal majoration of $\int\int_{R_i \times R_j} | x - y |^{-n} \, dx dy$, Lemma 3.4 will follow if we prove that $| < \tilde{T} 1_R, 1_R > | \leq C | R |$ for all cubes. Also, replacing the second 1_R by a smooth function Φ which is 1 on R and is supported on $2R$ only brings errors $\leq C | R |$ (the difference is $\leq C \int_R \int_{2R \setminus R} | x - y |^{-n} \leq C | R |$). So we only need to prove that

$$| < \tilde{T} 1_R, \Phi > | \leq C | R | \tag{32}$$

for such a Φ.

The idea is the following : if 1_R were smooth, then (32) would be nothing more than the weak boundedness of $\tilde{T}$. We'll show that the singularity of 1_R is not bad enough to destroy that estimate.

More precisely, we'll use the fact that the singularity lives on a small set (namely ∂R). We'll write 1_R as a sum of smooth functions, and show that the series giving (32) converges.

Let φ be a smooth function, supported on $B(0,r)$, where $r = | R |^{1/2}$, and with integral 1. We can choose φ so that

$$| \partial^\alpha \varphi | \leq C_\alpha r^{-n-|\alpha|} \tag{33}$$

for any multi-index α.

Write $1_R = \sum_{m \geq 0} f_m$, where $f_0 = 1_R * \varphi$, and then, for $m \geq 1$, $f_m = 1_R * \{2^{mn}\varphi(2^m x) - 2^{(m-1)n}\varphi(2^{m-1}x)\} = 1_R * \psi_m$, where ψ_m has integral 0 and is supported in $B(0, 2^{-m+1}r)$.

Next, cut each f_m into many pieces, using a partition of unity composed of functions supported in balls of radius $2^{-m}r$. We obtain $f_m = \sum_{k \in I(m)} g_{m,k}$, where each $g_{m,k}$ is supported in a ball of radius $2^{-m}r$. We can do this with a set of indices $I(m)$ such that

$$| I(m) | \leq C\, 2^{-m(n-1)}, \tag{34}$$

because f_m is supported on $\{x \ :\ \mathrm{dist}(x, \partial R) \leq 2^{-m+1}r\}$. Also, we can manage so that each $g_{m,k}$ satisfy

$$\| \partial^\alpha g_{m,k} \|_\infty \leq C_\alpha (2^{-m}r)^{-|\alpha|} \quad \text{for all} \quad \alpha \in \mathbb{N}^n. \tag{35}$$

Now, let us estimate each $< \tilde{T} g_{m,k}, \Phi >$. We choose a point x_0 in the support of $g_{m,k}$, and write $\Phi = \Phi_1 + \Phi_2$, where Φ_1 is supported in $B(x_0, 10r2^{-m})$, and Φ_2 is zero on $B(x_0, 5r2^{-m})$. We can easily manage to have $| \Phi_2 | \leq C$, and $\| \partial^\alpha \Phi_1 \|_\infty \leq C(2^{-m}r)^{-|\alpha|}$ for all α.

Applying the weak boundedness of $\tilde{T}$, we get $|< \tilde{T} g_{m,k}, \Phi_1 >| \leq C 2^{-mn} r^n$. On the other hand, $|< T g_{m,k}, \Phi_2 >| \leq C \int\int_D | x - y |^{-n}\, dy\, dx$, where $D = \{(x,y) \ :\ | y - x_0 | < 2^{-m+1}r,\ | x - x_0 | \geq 5r2^{-m}$ and $x \in 2R\}$. This is less than $C2^{-mn}\mathrm{Log}(m+1)r^n$, and so $|< T g_{m,k}, \Phi >| \leq C\, 2^{-mn}\mathrm{Log}(m+1)r^n$. Summing on k (using (34)) and then on m gives $|< T 1_R, \Phi >| \leq C r^n = C | R |$. This proves Lemma 3.4 and, by the same token, Lemma 3.3.

Remark. Lemma 3.4 seems a little long to prove (although the proof does not require too much thought), but it seems that one very often has to prove something like this. For instance, Y. Meyer's "commutation lemma" (see [My1]) is proved very much like this, and its use seems unavoidable in some instances. The usual way to solve the problem is of course to leave the easy proof as an exercise.

The estimates of Lemma 3.3 will be enough to imply that T is bounded. We want to apply Shur's lemma, but we cannot sum our estimates on $| C_{Q,R} |$ directly, because there are many more small cubes than large ones. So we'll sum the $|< \tilde{T} h_Q^\epsilon, h_{Q'}^{\epsilon'} >|$, for each Q and ϵ, against an appropriate weight that depends on the size of Q'. Let us first take sums on sets of cubes of the same size. Let Q be a given cube, with sidelength r. Call $\mathcal{A}_j(Q)$, $j \in \mathbb{Z}$, the set of all dyadic cubes with sidelength equal to $2^j r$.

Lemma 3.5.

$$\sum_{R \in \mathcal{A}_j(Q)} (| C_{Q,R} | + | C_{R,Q} |) \leq \begin{cases} C 2^{-jn/2} \; if \; j \geq 0 \\ C 2^{-jn/2} 2^{\delta j} \; if \; j < 0 \; and \; \delta \neq 1 \\ C 2^{-jn/2} \mid j \mid 2^j \; if \; j < 0 \; but \; \delta = 1. \end{cases} \tag{36}$$

Let us just prove the estimate for $| C_{Q,R} |$; the other one will follow by symmetry. First suppose that $j \geq 0$ (so that $| R | \geq | Q |$).

There are only a finite number of R's such that $\text{dist}(Q,R) \leq | R |^{1/n}$ and for these R's, (28) gives $| C_{Q,R} | \leq C | Q |^{1/2} | R |^{-1/2} \leq C 2^{-jn/2}$. For the other cubes R (those which are at distance $\geq 2^j r$ from Q), note that for a given $k \geq 0$, the number of cubes R such that $\text{dist}(Q,R) \sim 2^k 2^j r$ is $\leq C 2^{nk}$. For each of these cubes, $| C_{Q,R} | \leq C | Q |^{1/2} | R |^{1/2} r^\delta (2^k 2^j r)^{-n-\delta} \leq C 2^{-k(n+\delta)} 2^{-j(\frac{n}{2}+\delta)} \leq C 2^{-k(n+\delta)} 2^{-jn/2}$.

We now sum on all these cubes, and then on k, and get less than $C 2^{-jn/2}$. So (36) is valid in this case.

Let us now deal with the case when $j < 0$ (so that $| Q | > | R |$). First consider the cubes R that are at distance $\geq r$ from Q. For a given $k \geq 0$, the number of R's such that $2^k r \leq \text{dist}(Q,R) \leq 2^{k+1} r$ is less than $C 2^{nk} \frac{|Q|}{|R|} = C 2^{nk} 2^{-nj}$. For each of these cubes, $| C_{Q,R} | \leq C | Q |^{1/2} | R |^{1/2} | R |^{\delta/n} [2^k r]^{-n-\delta} \leq C 2^{-k(n+\delta)} 2^{j(\frac{n}{2}+\delta)}$. Summing on all these cubes, and then on k gives less than $2^{-j(\frac{n}{2}-\delta)}$, which is compatible with (36).

We are left with the cubes R such that $\text{dist}(Q,R) \leq r$. For each $0 \leq k \leq -j$, consider the set $\mathcal{A}_{j,k}$ composed of all the cubes R such that $2^k 2^j r \leq \text{dist}(R, \bigcup_\eta \partial Q_\eta) < 2^{k+1} 2^j r$ (for $k = 0$, let us even include the cubes that are at distance $\leq 2^j r$ from $\bigcup_\eta \partial Q_\eta$). To estimate the number of elements of $\mathcal{A}_{j,k}$, note that the volume they cover is $\leq C r^{n-1} (2^{k+1} 2^j r)$ (the surface of $\bigcup_\eta \partial Q_\eta$ is $\leq C r^{n-1}$, and we have to multiply by the approximate width). Consequently, $\# \mathcal{A}_{j,k} \leq C 2^k 2^j 2^{-nj}$. For each cube $R \in \mathcal{A}_{j,k}$,

$$\begin{aligned} | C_{Q,R} | &\leq C | R |^{1/2} | Q |^{-1/2} | R |^{\delta/n} (2^k 2^j r)^{-\delta} \\ &\leq C 2^{-k\delta} 2^{jn/2}. \end{aligned}$$

Summing on $\mathcal{A}_{j,k}$, we obtain less than $C 2^{k(1-\delta)} 2^j 2^{-jn/2}$. We can now sum on k. If $\delta = 1$, we have $-j-1$ equal terms of $2^j 2^{-jn/2}$, which gives (36) ; if $\delta < 1$, we get less than $C 2^{-j(1-\delta)} 2^j 2^{-jn/2}$, which again gives (36). This completes the proof of Lemma 3.5.

To apply Shur's lemma, let us compute the matrix of the operator TM_b in the basis (h_Q^ϵ) of the previous section. By (22), the coefficient of coordinates (Q,ϵ) and (Q',ϵ') of this matrix is $< TM_b h_Q^\epsilon, h_{Q'}^{\epsilon'} >_b = < \tilde{T} h_Q^\epsilon, h_{Q'}^{\epsilon'} > = C_{Q,Q'}^{\epsilon,\epsilon'}$. Also, by (23), TM_b is bounded on L^2 if and only if the matrix $\left(\left(C_{Q,Q'}^{\epsilon,\epsilon'} \right) \right)$ defines a bounded operator on $\ell^2(I)$, where I is the set of couples (Q,ϵ). As we said earlier, rather than compute $\sum_I | C_{Q,Q'}^{\epsilon,\epsilon'} |$, we want to sum against an appropriate weight.

To each $(Q,\epsilon) \in I$, let us associate the positive number $\omega(Q,\epsilon) = | Q |^{\frac{1}{2}-\frac{\delta}{2n}}$.

Lemma 3.6. *There is a constant $C \geq 0$ such that*

$$\sum_{(Q',\epsilon')\in I} | C_{Q,Q'}^{\epsilon,\epsilon'} | \omega(Q',\epsilon') \leq C\omega(Q,\epsilon) \quad for\ each \quad (Q,\epsilon) \tag{37}$$

and

$$\sum_{(Q,\epsilon)\in I} | C_{Q,Q'}^{\epsilon,\epsilon'} | \omega(Q,\epsilon) \leq C\omega(Q',\epsilon') \quad for\ each \quad (Q',\epsilon'). \tag{38}$$

This is easily obtained from Lemma 3.5. To prove (37), let us first choose $j \geq 0$ and sum on all cubes $Q' \in \mathcal{A}_j(Q)$ and all ϵ' (remember that, for each Q', there are less than C indices ϵ'). We get less than $C\{2^{nj}r^n\}^{\frac{1}{2}-\frac{\delta}{2n}}2^{-jn/2} \leq C \mid Q \mid^{\frac{1}{2}-\frac{\delta}{2n}} 2^{-j\delta/2}$. Summing on all $j \geq 0$ gives less than $C\omega(Q,\epsilon)$.

Next, pick $j < 0$, and sum on (Q',ϵ'), where $Q' \in \mathcal{A}_j(Q)$. If $\delta \neq 1$, we get $\leq C\{2^{nj}r^n\}^{\frac{1}{2}-\frac{\delta}{2n}}2^{-jn/2}2^{\delta j} \leq C \mid Q \mid^{\frac{1}{2}-\frac{\delta}{2n}} 2^{j\delta/2}$, and summing on j still gives less than $C\omega(Q,\epsilon)$. This remains true with trivial modifications when $\delta = 1$, and so (37) is true. Lemma 3.6 follows because our estimates on the $C_{Q,Q'}^{\epsilon,\epsilon'}$'s are symmetric.

It immediatly follows from Shur's lemma and Lemma 3.6 that TM_b (and therefore T) is bounded on $L^2(\mathbb{R})$. Let us even give a proof of Shur's criterion for the convenience of the reader.

Lemma 3.7. (Shur's criterion). *We are given a set of indices I and, for each $i \in I$, a number $\omega_i > 0$. Suppose that, for some $C \geq 0$, the matrix $((C_{i,j}))$ satisfies*

$$\sum_j | C_{i,j} | \omega_j \leq C\omega_i \quad for\ each\ i \tag{39}$$

and

$$\sum_i | C_{i,j} | \omega_i \leq C\omega_j \quad for\ each\ j. \tag{40}$$

Then the matrix $((C_{i,j}))$ defines a bounded operator on $\ell^2(I)$.

Proof. Given a vector $x = (x_j)$, we want to estimate $\| y \|$, where y is the vector with coordinates $y_i = \sum_j C_{ij}x_j$. We write $y_i = \sum_j \left(C_{i,j}^{1/2}\omega_j^{1/2}\right)\left(\omega_j^{-1/2}C_{i,j}^{1/2}x_j\right)$ and apply Schwarz :

$$| y_i |^2 \leq \left\{\sum_j | C_{i,j} | \omega_j\right\}\left\{\sum_j | C_{i,j} | \omega_j^{-1} | x_j |^2\right\} \leq C\omega_i \sum_j | C_{i,j} | \omega_j^{-1} | x_j |^2$$

by (39).

Then $\| y \|^2 \leq C \sum_j \sum_i \omega_i \mid C_{i,j} \mid \omega_j^{-1} \mid x_j \mid^2 \leq C \sum_j \omega_j \omega_j^{-1} \mid x_j \mid^2 = C \| x \|^2$, which proves the lemma !

The proof of Theorem 1.11 is now complete in the special case when $Tb = T^t b = 0$ for some paraaccretive function b. Let us say how the argument above should be modified when $Tb_1 = T^t b_2 = 0$ for two different paraaccretive functions. One considers two different bases $(h_Q^{\epsilon,1})$, $(Q,\epsilon) \in I_1$ and $(h_Q^{\epsilon,2})$, $(Q,\epsilon) \in I_2$ (the two sets of indices could be different if b_1 and b_2 are not "special paraaccretive"), the first one adapted to b_1, and the second one adapted to b_2. Then, one computes the coefficients of TM_{b_1} in these two bases. One gets the numbers $< T(b_1 h_Q^{\epsilon,1}), h_{Q'}^{\epsilon',2} >_{b_2} = < \tilde{T} h_Q^{\epsilon,1}, h_Q^{\epsilon,2} >$, where $\tilde{T} = M_{b_2} T M_{b_1}$. The computations are then done exactly as above.

4. End of the proof : paraproducts

To complete the proof of the Tb-theorem, we shall build some sort of a paraproduct. Here, too, ideas related to wavelets will help us : we shall use a simple modification of the paraproduct using wavelets defined by Y. Meyer.

To simplify notations, we shall still consider the case when $b_1 = b_2 = b$; the general case would only require minor modifications. Let us first define the equivalent of the function φ of Part I. For each Q, call

(41) $$\theta_Q(x) = \left\{ \int_Q b(t)\,dt \right\}^{-1} 1_Q(x).$$

Note that $\| \theta_Q \|_\infty \leq C \mid Q \mid^{-1}$, and that $\int \theta_Q b = 1$; if b is not "special-paraaccretive", Q runs through the modified cubes of the end of Section 2.

Next, let us recall how one proves that the "wavelet coefficients" of a function of BMO satisfy a Carleson measure condition.

Lemma 4.1. *If $\beta \in BMO$, and if the $C_Q^\epsilon = \int \beta(x) h_Q^\epsilon(x) b(x)\,dx$ are the coefficients of β in the basis (h_Q^ϵ), then, for each cube R,*

(42) $$\sum_Q \sum_\epsilon \mid C_Q^\epsilon \mid^2 \leq C \mid R \mid,$$

where the sum is taken over all (Q,ϵ) such that $Q \subset R$.

The (very classical) proof is similar to our earlier proof of the square function estimate. Note that $\int \beta(x) h_Q^\epsilon(x) b(x)\,dx$, fortunately, does not change when a constant is added to β. If $Q \subset R$, then

$$\begin{aligned} C_Q^\epsilon &= \int \beta(x) 1_R(x) h_Q^\epsilon(x) b(x)\,dx \\ &= \int [\beta(x) - m_R\beta] 1_R(x) h_Q^\epsilon(x) b(x)\,dx = < (\beta - m_R\beta) 1_R, h_Q^\epsilon >_b . \end{aligned}$$

Consequently, it follows from (23) that

$$\sum_{Q\subset R}\sum_{\epsilon} | C_Q^\epsilon |^2 \leq C \parallel (\beta - m_R\beta)1_R \parallel_2^2 \leq C \mid R \mid,$$

and so (42) is true.

Next, for each sequence C_Q^ϵ satisfying (42), we shall define an operator $P\left((C_Q^\epsilon)\right) = P$ by its kernel

$$P(x,y) = \sum_{Q}\sum_{\epsilon} C_Q^\epsilon h_Q^\epsilon(x)\theta_Q(y). \tag{43}$$

Lemma 4.2. $| P(x,y) | \leq C \mid x-y \mid^{-n}$.

Proof. Let $x \neq y$ be given. Note that $h_Q^\epsilon(x)\theta_Q(y) \neq 0$ implies that x and y are in Q. Then the size of Q is larger than $| x-y | /C$ and also, for each size $2^k \geq | x-y | /C$, there is at most one cube Q of size 2^k for which $h_Q^\epsilon(x)\theta_Q(y) \neq 0$. Summing the absolute values gives less than $\sum_k\sum_\epsilon C \mid C_Q^\epsilon \mid 2^{-kn/2}2^{-kn} \leq C\sum_k 2^{kn/2}2^{-3kn/2}$ (because (42) implies that $| C_Q^\epsilon | \leq C \mid Q \mid^{1/2}$) $\leq C\sum_k 2^{-kn} \leq C \mid x-y \mid^{-n}$, as needed.

Next, we wish to show that (43) defines a bounded operator from $b\mathcal{D}$ to its dual. This is because, if f and g are test-functions, one can show that the series

$$\sum_Q\sum_\epsilon | C_Q^\epsilon | \left| \int\int g(x)b(x)h_Q^\epsilon(x)\theta_Q(y)b(y)f(y)dx\,dy \right|$$

converges. Here is why : if Q is a large cube, the corresponding terms are quite small because f and g have a fixed compact support, and the L^∞–norms of θ_Q and h_Q^ϵ tend to zero fast enough ; when Q is small, one uses the smoothness of g and the fact that $\int g(x_0)b(x)h_Q^\epsilon(x) = 0$ if x_0 is any point of Q. So P is well-defined.

Now, we want some weak smoothness property for the kernel ($P(x,y)$ does not satisfy (3) because of the jumps of h_Q^ϵ and θ_Q).

Lemma 4.3. *Let Q be a dyadic cube and h a bounded function, supported on Q and such that $\int h(x)b(x)dx = 0$. Then, for all $x \notin 2Q$, $P(bh)(x) = P^t(bh)(x) = 0$.*

Note that this estimate is much better than the conclusion of Lemma 3.1. To prove Lemma 4.3, note that $P(bh)(x)$ is a sum of terms of the form $C_{Q'}^\epsilon h_{Q'}^\epsilon(x)\int \theta_{Q'}(y)h(y)b(y)dy$, and all these terms are zero, unless Q' contains x, and also some point $y \in Q$. Since $x \notin Q$, Q' contains Q strictly. Then $\theta_{Q'}$ is constant on Q, and the integral is zero, so $P(bh)(x) = 0$ for all $x \notin Q$.

For the transpose, we write $P^t(bh)(x)$ as a sum of terms of the form

$$C^{\epsilon}_{Q'}\theta_{Q'}(x)\int h^{\epsilon}_{Q'}(y)h(y)b(y)dy\ ;$$

as before, Q' must contain Q strictly, and then $h^{\epsilon}_{Q'}(y)$ is constant on Q. Altogether, $P^t(bh)(x) = 0$.

Lemma 4.4. *If we choose $C^{\epsilon}_Q = \int \beta(x)h^{\epsilon}_Q(x)b(x)dx$, then $Pb = \beta$ and $P^t b = 0$.*

Let us say what we mean by this. We want to show that, if R is a large cube tending to ∞ (for instance, take for R the union of the 2^n dyadic cubes of size 2^M touching 0 and let $M \to +\infty$), then $< P(b1_R), bg >$ tends to $\int \beta(x)b(x)g(x)$ for each $g \in \mathcal{D}$ such that $\int bg = 0$. A similar meaning is given to $P^t b$.

Note that if T is a SIO, and $Tb = \beta$, then $< T(b1_R), bg >$ tends to $\int \beta(x)b(x)g(x)$, too, for each $g \in \mathcal{D}$ such that $\int bg = 0$. This can easily be checked from the definition of Tb (Remark 1 following Theorem 1.11).

To prove Lemma 4.4, we shall do the computation formally, and leave the details of the limiting argument (involving the large cube R) to the reader. First,

$$P^t b(y) = \int \sum_Q \sum_{\epsilon} C^{\epsilon}_Q h^{\epsilon}_Q(x)\theta_Q(y)b(x)dx = 0$$

because each $\int h^{\epsilon}_Q(x)b(x)dx$ is zero. Also,

$$Pb(x) = \sum_Q \sum_{\epsilon} C^{\epsilon}_Q h^{\epsilon}_Q(x)\int \theta_Q(y)b(y)dy = \sum_Q \sum_{\epsilon} C^{\epsilon}_Q h^{\epsilon}_Q(x)$$

(by definition of θ_Q). This is $\beta(x)$ by definition of the C^{ϵ}_Q's, and (22).

We also need to show that P is bounded.

Lemma 4.5. *P is bounded on L^2, with a norm $\leq C \parallel \beta \parallel_{BMO}$.*

Proof. If $f \in b\mathcal{D}$,

$$Pf = \sum_Q \sum_{\epsilon} C^{\epsilon}_Q h^{\epsilon}_Q \int \theta_Q(y)f(y)dy,$$

and so, by (23),

$$\parallel Pf \parallel_2^2 \leq C \sum_Q \sum_{\epsilon} | C^{\epsilon}_Q |^2 \left| \int \theta_Q(y)f(y)dy \right|^2$$

$$\leq C \sum_Q \sum_{\epsilon} | C^{\epsilon}_Q |^2 \left\{ \frac{1}{| Q |}\int_Q | f(y) | \, dy \right\}^2 \leq C \parallel f \parallel_2^2,$$

by Carleson's theorem, because the $| C_Q^\epsilon |^2$ satisfy the Carleson measure condition (42) (again see [Ga3], Th. I.5.6).

Let us now put everything together. Let T be like in the theorem, and suppose that $Tb = \beta_1 \in BMO$ and $T^t b = \beta_2 \in BMO$. Construct operators P_1 and P_2 with the functions β_1 and β_2, and consider the difference $T' = T - P_1 - P_2^t$.

Since each piece does, T' is associated to a kernel that satisfies (2), and it also satisfies the conclusion of Lemma 3.1 (use Lemma 4.3).

Next, $M_b T' M_b$ is weakly bounded, because $M_b(P_1 + P_2^t)M_b$ is bounded and $M_b T M_b$ is weakly bounded.

Finally, Lemma 4.4 implies that if R is a large cube tending to ∞ (like in Lemma 4.4), then $< T'(b1_R), bg >$ tends to 0 when g is a test function such that $\int g(x)b(x) = 0$. A similar estimate is true for T'^t, and so we can apply the proof of Section 3, and obtain that T' is bounded.

Since P_1 and P_2 are bounded, we get the boundedness of T by substraction.

If the functions b_1 and b_2 were different, one would have to define the paraproducts a little differently. For instance, the kernel of P_1 would be

$$P_1(x,y) = \sum_{(Q,\epsilon)} \sum_{\in I} C_Q^\epsilon h_Q^{\epsilon,2}(x)\theta_Q(y),$$

where the set of cubes Q (replacing the dyadic cubes when b_1 or b_2 is not "special paraaccretive") is adapted both to b_1 and b_2 [the construction of such a set is just an iteration of the end of Section 2]. The rest of the proof is the same.

We finally completed the proof of Theorem 1.11.

5. Comments on Tb, spaces of homogeneous type.

Let us start with a comment on the proof. A first proof of (part of) the Tb-theorem using wavelets was given by Tchamitchian [Tc3]. However, this proof was not quite as simple. The proof given above is quite stricking, because it gives directly the "full" Tb-theorem. A first glance at the number of pages could make the reader mistakenly think that it is not so simple, after all. Let us suggest to the reader that would be tempted by such a thought to go and have a look at [DJS] ! Also, we did our best to give all the relevant details about the proof : it would be much shorter if we had restricted ourselves to the case when $b_1 = b_2 = b$ is accretive and defined on the real line. For a compact proof, see [CJS].

Another nice feature about this proof is that it extends rather easily to spaces of homogeneous type. We do not wish to give a detailed explanation, and so we'll not even say what a space of homogeneous type is. We refer to [CW], [McS1] and [McS2] for this. The point is that the notion of a singular integral operator extends nicely to such a space (see [McS1& 2]) and there is even an extension of the Tb-theorem to the case of spaces of homogeneous type. If we look at the proof given in Sections 2, 3 and 4, we see that we only used the existence on $\mathbb{R}^n$ of something like dyadic cubes. If the space of homogeneous type E has a family $\mathcal{R}_k$, $k \in \mathbb{Z}$, of partitions of E with the properties below, the proof given in the previous sections will extend to give a Tb-theorem on $L^2(E)$: we used the fact that

- if $Q \in \mathcal{R}_k$ and $Q' \in \mathcal{R}_{k'}$ for a $k \leq k'$, then Q contains Q' as soon as $Q \cap Q' \neq \emptyset$;

- all the cubes in $\mathcal{R}_k$ have comparable mass : $Q, Q' \in \mathcal{R}_k$ implies that $| Q | \leq C | Q' |$;

- each cube of $\mathcal{R}_k$ contains $\leq C$ cubes of $\mathcal{R}_{k+1}$;

- each cube $Q \in \mathcal{R}_k$ has a "sufficiently small boundary".

We used the last condition in Lemma 3.4, to make sure that 1_Q would be regular enough to allow the use of the weak boundedness, and in Lemma 3.6 to control the number of small cubes R that are close to $\bigcup_\eta \partial Q_\eta$. A condition that would be enough, for instance, would be the existence of an $\alpha > 0$ such that, if Q is one of the cube and r is its diameter, then

$$| \{x \ : \ \mathrm{dist}(x, \partial Q) \leq 2^{-k} r\} | \leq C 2^{-k\alpha} | Q | .$$

The existence of such cubes is not hard to establish in the special case of an Ahlfors-regular subset of some $\mathbb{R}^n$, i.e. a closed subset E of $\mathbb{R}^n$ with a measure μ (generally the restriction to E of the k–dimensional Hausdorff measure) such that, for some positive real number $k > 0$ and some $C \geq 0$, $C^{-1} r^k \leq \mu(E \cap B(x,r)) \leq C r^k$ for all $x \in E$ and $r > 0$.

A construction is given in [Dv6], but the author is not very proud of it (it is more complicated than necessary), and since it is used in a few different places in this book, a slightly simplified version will be added as an appendix.

What about all the other spaces of homogeneous type, then ? It turns out that another construction can be given, and that this construction extends to all spaces of homogeneous type (see [Ch.2]) ! Moreover, M. Christ's construction is not much more complicated. He uses it to prove a variant of the Tb-theorem, but with variable b (see [Ch.2], but also [Ch. 1] for a statement of the result).

A final comment on this proof : the fact that we do not have to deal with complicated approximations of the identity makes it easier to extend the theorem to the case of matrix-valued kernels. You will see an example of this in Part III, Theorem 6.7.

Let us not insist more.

6. Applications

We'll make this chapter a little shorter than it should be. The kind reader is referred to [My4] and [Ch.1] for most applications.

A. More wavelets

We start with two results at the interface between singular integrals and wavelets.

Theorem 6.1. *Let ψ be a C^1–function on $\mathbb{R}$, with rapid decay at ∞, and such that the $2^{j/2}\psi(2^j x - k)$, $(j,k) \in \mathbb{Z}^2$, form an orthonormal basis of $L^2(\mathbb{R})$ (for instance, take Y. Meyer's wavelet). Then the $2^j\psi(2^j x - k)$ form an unconditional basis of the atomic space $H^1(\mathbb{R})$.*

Proof. One needs to show that if $f \in H^1$ is a finite sum of the form

$$\sum_{j,k} C_{j,k}[2^j\psi(2^j x - k)] = \sum_I C_I\psi_I$$

(with a self-explanatory change of notations), and if (ϵ_I) is a sequence of complex numbers of modulus ≤ 1, then $f_\epsilon = \sum_I \epsilon_I C_I \psi_I$ satisfies $\| f_\epsilon \|_{H^1} \leq C \| f \|_{H^1}$ [with a C that does not depend on f or (ϵ_I) !].

Call T_ϵ the operator which sends ψ_I to $\epsilon_I\psi_I$. T_ϵ is well defined on L^2, and has a norm ≤ 1 because the $2^{j/2}\psi(2^j x - k)$ are an orthonormal basis of L^2. The kernel of T_ϵ is $\sum_{j,k} 2^{-j}\epsilon_{j,k}\psi(2^j x - k)\,\overline{\psi(2^j y - k)}$, and so one checks easily that T_ϵ is a singular integral operator. Finally, $T_\epsilon^t 1 = 0$, and it follows from the Remark 3 after Theorem 1.6 that the T_ϵ's are uniformly bounded on H^1. (Hence we proved the theorem.) We leave the estimates on the kernel as an easy exercise.

Remark. We stated this theorem because it is rather easy ; the same sort of proof would have shown that the ψ_I's are an unconditional basis for many other spaces (for H^1, however, the existence of such a basis is not so recent). Theorem 6.1 also generalizes to H^1 of the bi-disk (the idea is to use the theory of Calderón-Zygmund operators on product spaces). See [Le].

Our second result is a counterexample due to Lemarié (a first example was found, shortly before, by P. Tchamitchian [Tc1]). The question was the following : if T is a bounded singular integral operator, and is invertible on L^2, is it true that it is also invertible on L^p, $1 < p < +\infty$? The answer is no : for each p, there is an operator T like that which is not invertible on L^p.

Note that one cannot hope a better counterexample : an easy use of complex interpolation shows that if T is bounded on L^p for $p_0 \leq p \leq 4p_0^{-1}$ for some $p_0 < 2$, and is invertible on L^2, then it is also invertible on L^p for p in a small neighborhood of 2 (which of course depends on the norms of T on the L^p's). This was observed by Calderón [Ca2].

Here is the counterexample. Consider Y. Meyer's basis of $L^2(\mathbb{R})$ (for instance) $2^{j/2}\psi(2^j x - k)$, and index it by dyadic intervals $I = [k2^{-j}, (k+1)2^{-j}]$, so as to get a basis $(\psi_I)_{I \text{ dyadic}}$. If I is a dyadic interval, call $\Phi(I) = [0, 2^{j+1}]$ if $I = [0, 2^j]$ and $\Phi(I) = \emptyset$ otherwise.

Now let T_0 be the only operator such that $T_0\psi_I = 0$ if $\Phi(I) = \emptyset$, and $T_0\psi_I = \psi_{\Phi(I)}$ otherwise. Clearly, T_0 is a bounded operator on L^2, with norm 1. Also, the kernel of T_0 is

$\sum_j \psi_{[0,2^{j+1}]}(x)\bar{\psi}_{[0,2^j]}(y)$; it satisfies (2) and (3) by the usual computation, and so T_0 is a SIO.

Thus, if $0 < r < 1$, the operator $T = 1 - rT_0$ is a SIO, and is invertible on $L^2(\mathbb{R})$. The inverse is $T^{-1} = \sum_{k\geq 0} r^k T_0^k$, and in particular

$$T^{-1}(\psi_{[0,1]}) = \sum_{k\geq 0} r^k \psi_{\Phi^k([0,1])} = \sum_{k\geq 0} r^k 2^{-k/2}\psi(2^{-k}x).$$

We leave the fact that, given a $p < 2$, one can find $r < 1$ such that this function is not in L^p as an exercise (one can do it directly, or use the characterization of L^p by wavelet coefficients given in Part I, § 8, Example 2).

If one wishes a counterexample for $p > 2$, one can use the transposed operator (which corresponds to taking $\Phi([0,2^k]) = [0,2^{k-1}])$.

In spite of this example, it is possible to prove positive results concerning the inversion of singular integral operators. See [Tc4], for instance.

B. The Cauchy integral and related operators.

In most of the following examples, the operators will be principal value operators defined by an antisymmetric kernel (Example 1.3). For these operators, we said in Section 1 that the weak boundedness property is always satisfied. Since $T^t 1 = -T1$, one will only have to check that $T1 \in BMO$ to apply the T1-theorem (a similar remark can be formulated concerning the Tb-theorem).

Example 6.2. (Calderón's commutators). Let A be a Lipschitz function on the real line (i.e., a function such that $|A(x) - A(y)| \leq C | x - y |$). Set $K_n(x,y) = \frac{(A(x)-A(y))^n}{(x-y)^{n+1}}$, and let T_n be the principal valeur operator defined by K_n.

Exercise : prove that T_n is bounded on $L^2(\mathbb{R}^n)$, with a norm $\leq C^{n+1} \| A' \|_\infty^n$ for some $C \geq 0$.

Hint : do this by induction. After checking that K satisfies (2) and (3) with $C_0 \leq C(n+1) \| A' \|_\infty^n$, show that $T_n(1) = T_{n-1}(A')$ and conclude.

Comments. The first proof of this estimate dates from Calderón [Ca1], in 1977. In their famous paper [CMM], Coifman, McIntosh and Meyer proved a polynomial estimate, namely $\| T_n \| \leq C(1+n^4) \| A' \|_\infty^n$; this estimate was then improved by Murai [Mu1], by a completely different technique. The best estimate, up to now, is due to Christ and Journé : for each $\epsilon > 0$,

(44) $$\| T_n \| \leq C_\epsilon (1 + \| A' \|_\infty)^{1+\epsilon} \quad \text{(see [ChJ]).}$$

Example 6.3. (The Cauchy integral on a Lipschitz graph).

Let $A : \mathbb{R} \to \mathbb{R}$ be Lipschitz. The antisymmetric standard kernel $K(x,y) = [x + iA(x) - y - iA(y)]^{-1}$ defines a principal value operator C_A.

Exercise

- Write

$$K(x,y) = \frac{1}{x-y}\left[1 + i\frac{A(x)-A(y)}{x-y}\right]^{-1}$$

$$= \sum_n (-i)^n \frac{1}{x-y}\left(\frac{A(x)-A(y)}{x-y}\right)^n = \sum_n (-i)^n K_n(x,y) \ ;$$

deduce Calderón's result from the estimate of Example 6.2 : C_A is bounded as soon as $\| A' \|_\infty$ is small enough.

- Prove Coifman, McIntosh and Meyer's result directly : C_A is bounded for all Lipschitz A's (Hint : compute $C_A(1 + iA')$).

Comments. The first proof of the boundedness of C_A when $\| A' \|_\infty$ is small did not use commutators. Calderón [Cal] showed that the $\| C_{tA} \|$, where A is given and $0 \le t \le 1$, satisfy a differential relation which allowed him to estimate C_A ; the estimates on commutators are a consequence of the boundedness of C_A.

Since the first proof of the boundedness of C_A in the general case [CMM], it has become fashionable to give new proofs of the boundedness of C_A. The reader is invited to produce one or two of his own. For a nice collection, see T. Murai's book [Mu2] ; for the "shortest" proof, see [CJS] (it is also in [Mu2]).

The best estimate on the norm of C_A is

$$\| C_A \| \le C(1 + \| A' \|_\infty)^{1/2}. \tag{45}$$

It is due to T. Murai [Mu1] (also see [Mu2]) ; the proof uses a perturbation argument similar to the one presented in the first sections of Part III (but sharper !). The estimate (45) is quite good, because one can find functions A, with $\| A' \|_\infty \to +\infty$, such that $\| C_A \| \ge 10^{-4} \| A' \|_\infty^{1/2}$. The idea is to approximate Garnett's example (see Part III) by a lipschitz graph [Dv3].

Note that it is not a bad idea to have a very good estimate for C_A, because many other operators can be constructed from C_A. See Example 6.5 below, for instance.

Example 6.4. (Chord-arc curves). A rectifiable, connected curve $\Gamma \subset \mathbb{C}$ is called chord-arc if, for some constant $C \ge 0$ and any two points A , $B \in \Gamma$, the length of the piece of Γ connecting A to B is less than C times $| B - A |$. (This is for an open curve Γ ; a small modification, left to the reader, has to be done for closed curves.)

Let Γ be a chord-arc curve, and denote by ds the arc-length measure on Γ. One can try to define a Cauchy integral on Γ by $C_\Gamma f(x) = \text{p.v.}\int_\Gamma \frac{1}{z-w} f(w)ds(w)$ for $f \in L^2(\Gamma, ds)$.

Exercise. Show that C_Γ is a bounded operator on $L^2(\Gamma, ds)$. (Do not pay too much attention to the definition of the principal value.) Hint : call $t \to z(t)$ a parametrization of Γ by arclength. The problem is equivalent to showing that $K(x,y) = [z(x) - z(y)]^{-1}$ defines a bounded operator T on $L^2(\mathbb{R})$, and this is easily done because $Tz' = 0$ and z' is "special paraaccretive".

Example 6.5. Let $A : \mathbb{R} \to \mathbb{R}^N$ be a Lipschitz function, and let F be a C^∞ function from $\mathbb{R}^N$ to $\mathbb{C}$. The kernel $\frac{1}{x-y} F\left(\frac{A(x)-A(y)}{x-y}\right)$ defines a bounded singular integral operator on $L^2(\mathbb{R})$.

One can find a first proof of this fact in [CDM]. The idea is to write F as a Fourier series, and then to study the case of $F(x) = e^{i(x_1\xi_1+\cdots+x_N\xi_N)}$. One is trivialy reduced to the case of one dimension, and then one writes the exponential e^{ix} in terms of an integral of objects like $(1+ix)^{-1}$ (using the Cauchy formula). One gets an estimate on the operator with kernel $\frac{1}{x-y}\exp\left(i\frac{B(x)-B(y)}{x-y}\right)$ which grows polynomially in $\| B' \|_\infty$, and this estimate is enough to sum the operators coming from the Fourier series of F, provided that the Fourier coefficients have enough decay.

Of course, the condition " $F \in C^\infty$ " is not needed ; the better one can estimate the norm of $\mathcal{C}_A$, the less derivatives will be needed on F. So it is advisable to use Murai's result (45). For a reasonably sharp estimate, see [DS1].

Using this example, and the "rotation method" to get operators on $L^2(\mathbb{R}^n)$, one can get quite a collection of Calderón-Zygmund operators.

Example 6.6 ("Stable kernels"). Let $K(x,y)$ be a standard kernel on the line ; we'll say that K is a Calderón-Zygmund kernel if there is a bounded SIO such that K is its kernel. We'll say that K is "stable" if, for all Lipschitz functions $A : \mathbb{R} \to \mathbb{R}$, the kernel $K(x,y)\frac{A(x)-A(y)}{x-y}$ is a Calderón-Zygmund kernel (so, taking $A(x) = x$, K is Calderón-Zygmund).

It was first observed by Journé that if K is an antisymetric Calderón-Zygmund kernel, then it is stable.

Exercise : use the T1-theorem to prove it.

One can easily characterize the Calderón-Zygmund kernels that are stable. First, the notion only depends on the kernel, and can be studied on truncated kernels $K(x,y)1_{\{\epsilon \le |x-y| \le M\}}$. Thus one can restrict to compactly supported kernels (and prove uniform estimates). Then, one applies the T1-theorem and does an integration by parts, and one finds that the Calderón-Zygmund kernel K is stable if and only if the function β defined by

$$\beta(u) = \iint_{x<u<y} \frac{K(x,y)\,dx\,dy}{x-y} - \iint_{y<u<x} \frac{K(x,y)\,dx\,dy}{x-y} \tag{46}$$

is in $BMO(\mathbb{R})$.

Note that, if K is a convolution kernel, or if K is antisymmetric, the condition is automatically satisfied.

The amusing thing is that, if K is stable, then $K(x,y)\frac{A(x)-A(y)}{x-y}$ is also stable when A is Lipschitz, so that one can prove that the kernel $K(x,y)\left(\frac{A(x)-A(y)}{x-y}\right)^N$ also defines a bounded singular integral operator, with norm $\leq C(K)^{N+1} \| A' \|_\infty^N$. One can then build other bounded operators, like when $K(x,y) = \frac{1}{(x-y)}$, by summing series. This is a joint work with Coifman, Journé and Semmes, but we can only refer to [Dv2] for details. There was a partial result by Murai.

Example 6.7 (singular integrals on Lipschitz graphs). Let $0 < d \leq n$ be integers, and let $k(z)$ be a C^∞ function, defined on $\mathbb{R}^n \backslash \{0\}$, and such that

$$|\nabla^j k(x)| \leq C(j) \, |x|^{-d-j} \text{ for all } j \geq 0 \tag{47}$$

and

$$\sup_{0<\epsilon<M} \left| \int_{\epsilon<|t|<M} k(t\theta) \, |t|^{d-1} \, dt \right| \leq C \text{ for all } \theta \in S^{n-1}. \tag{48}$$

Let $A : \mathbb{R}^d \to \mathbb{R}^{n-d}$ be a Lipschitz function. Then the kernel

$$K(u,v) = k(u-v \, , \, A(u) - A(v))$$

defines a bounded singular integral operator on $L^2(\mathbb{R}^d)$.

The proof is a nice application of the techniques we have just seen (and, unfortunately, of techniques that will be seen in Part III). We shall write it as a series of exercises.

Exercises. Let $k : \mathbb{R}\backslash\{0\} \to \mathbb{C}$ be such that

$$|k(x)| \leq C_0 \, |x|^{-1} \text{ and } |k'(x)| \leq C_0 \, |x|^{-2}, \tag{49}$$

and also

$$\sup_{0<\epsilon<M} \left| \int_{\epsilon<|x|<M} k(x)dx \right| \leq C_0. \tag{50}$$

1. Show that $K(x,y) = k(x-y)$ is a Calderón-Zygmund kernel. You will not even need to apply the T1-theorem, because the Fourier transform of truncated kernels of the form $\chi(x)k(x)$ can be estimated directly.

2. Show that $k(x-y)$ is even a stable kernel, and that $K(x,y) = k(x-y)\left(\frac{A(x)-A(y)}{x-y}\right)^N$ defines a bounded operator on $L^2(\mathbb{R})$, with norm $\leq C^{N+1} C_0 \| A' \|_\infty^N$ for some C.

3. Show that, if A is Lipschitz, the kernel $k(x-y) \exp i \left(\frac{A(x)-A(y)}{x-y}\right)$ defines a bounded operator, with norm $\leq C(\| A' \|_\infty) C_0$.

The estimate above will not give a constant $C(\| A' \|_\infty)$ which is only polynomial in terms of $\| A' \|_\infty$. One now has to apply the technique of Part III, § 4.B, which shows that one can take $C(\| A' \|_\infty) \leq C(1+ \| A' \|_\infty)^M$ for some M. We therefore postpone this part of the exercise, and suppose we get a polynomial estimate in $\| A' \|_\infty$.

Now consider the kernel of Example 6.7, in the special case when $d = 1$. After a change of coordinates in $\mathbb{R}^n$ of the form $(x_1, \cdots, x_n) \to (x_1, Cx_2, \cdots Cx_n)$, we are reduced to the case when $\| A' \|_\infty \leq 1$. The values of $k(x) = k(u, x')$ (we shall systematically write $x = (u, x')$ with $u \in \mathbb{R}$ and $x' \in \mathbb{R}^{n-1}$) when $| x' | > | u |$ do not matter, and so we can modify them at will. Write $k(x) = \tilde{k}(u, \theta)$, where $\theta = \frac{x'}{u}$ has its values in $\mathbb{R}^{n-1}$. We can assume that this function is 2π-periodic (in all variables) in θ.

4. Check that $\tilde{k}(\cdot, \theta)$ is a C^∞ function of θ, valued in the Banach space of kernels satisfying (49) and (50). Thus, one can write

$$\tilde{k}(u, \theta) = \sum_{\nu \in \mathbb{Z}^n} k_\nu(u) e^{2i\pi\theta.\nu}, \tag{51}$$

where the kernels k_ν satisfy (49) and (50) with constants $C_0(\nu)$ that decrease rapidly as $| \nu | \to +\infty$.

5. Coming back to $K(u, v) = k(u - v, A(u) - A(v))$, use the fact that $K(u, v) = \sum_\nu k_\nu(u - v) \exp 2i\pi\nu. \left(\frac{A(u) - A(v)}{u - v}\right)$ to deduce the case $d = 1$ of Example 6.7.

6. Finally, deduce the general case from the case when $d = 1$ by the "method of rotations" : write truncated operators

$$T_\epsilon f(u) = \int_{|v-u|>\epsilon} K(u, v) f(v) dv = \int_{\theta \in S^{d-1}} T_{\theta,\epsilon} f(u) d\theta,$$

where

$$T_{\theta,\epsilon} f(u) = \int_{|t|>\epsilon} K(u, u + t\theta) f(u + t\theta) \; | t |^{d-1} \, dt$$

can be controlled by the 1-dimensional result. The reader who is not too familiar with the method of rotations will find descriptions in many sources, and a proof of the exercise can be found in [Dv5], p. 245.

Remark. When $k(x)$ is odd and homogeneous of degree $-d$, most of the exercise can be avoided : one can prove the result by a direct application of Example 6.5 and the method of rotations.

PART III

SINGULAR INTEGRALS ON CURVES AND SURFACES

—ooOoo—

1. Introduction and notations

We would like to extend Coifman, McIntosh and Meyer's nice result to as many higher-dimensional objects as possible. In the following, we'll be given $0 < k \leq n$, and a k-dimensional object $S \subset \mathbb{R}^n$. We'll always call S a "surface", but we do not assume any smoothness on S : S is a set. Also, S will come with a non negative Radon measure μ on $\mathbb{R}^n$, such that supp $\mu = S$. For instance, we could take for S a smooth k-dimensional surface, with for μ the k-dimensional surface measure on S.

The singular integrals we wish to consider are generalizations of the Cauchy integral (defined on curves of the complex planes).

Definition 1.1. The function $K(z)$, defined on $\mathbb{R}^n \setminus \{0\}$, will be called a "good kernel" if $K(-z) = -K(z)$ for all $z \neq 0$, K is C^∞ and

(1) $$|\nabla^j K(x)| \leq C(j) \, |x|^{-k-j} \quad \text{for all } j \geq 0.$$

Remark 1.2. The choice of the class of "good kernels" is not extremely important: we only have to take it small enough, so that defining singular integrals on lipschitz graphs will not be a problem. We could also have taken the slightly larger class where the antisymmetry condition is replaced by

(2) $$\sup_{0<\epsilon<M} \left| \int_{\epsilon<|t|<M} K(t\theta) \, |t|^{k-1} \, dt \right| \leq C \quad \text{for all } \theta \in S^{n-1}.$$

The cancellation still has to occur on each line, because we want cancellation on all k-planes, not matter how oriented.

We could also have asked for the more restrictive condition that K be C^∞, odd, and such that $K(\lambda x) = \lambda^{-k} K(x)$ for all $x \neq 0$ and $\lambda > 0$. This class would not be enough, however, for the current proof of the converse result mentioned in Section 9 (Theorem 9.5) to work.

Given a good kernel and a Radon measure $\mu \geq 0$ on $\mathbb{R}^n$, we would like to define an operator T by a formula like $Tf(x) = \int K(x-y) f(y) d\mu(y)$. We'll always assume that μ is in the following class Δ.

Definition 1.3. The letter Δ will denote the class of non negative Radon measures μ on $\mathbb{R}^n$ such that, for some $C \geq 0$,

(3) $$\mu(B(x,r)) \leq Cr^k \quad \text{for all } x \in \mathbb{R}^n \text{ and } r > 0.$$

We'll see soon that $\mu \in \Delta$ is necessary if we want all good kernels to define bounded operators on $L^2(\mathbb{R}^n, d\mu)$. Let us first define a truncated and a maximal operator. For $\epsilon > 0$ and $f \in C_c(\mathbb{R}^n)$, let

$$T_\mu^\epsilon f(x) = \int_{|x-y|>\epsilon} K(x-y)f(y)d\mu(y) \tag{4}$$

and

$$T_\mu^* f(x) = \sup_{\epsilon > 0} | T_\mu^\epsilon f(x) | . \tag{5}$$

If the maximal operator T_μ^* is bounded on $L^2(\mathbb{R}^n, d\mu)$, the T_μ^ϵ's are uniformly bounded, and we'll be able, if we want, to define an operator T by extracting a weakly convergent sequence. On the other hand, studying T_μ^* will not be harder than just showing the boundedness of some limit of T_μ^ϵ's ; in fact, we'll prove later that, with a small additional hypothesis on μ, the L^2–boundedness of such a limit implies the boundedness of T_μ^*.

Our main question, in its general form, is the following : for which measures μ is T_μ^* bounded on $L^2(\mathbb{R}^n, d\mu)$? When $k = 1$ and $n = 2$, we could consider the special case of $K(z) = \frac{1}{z}$, and ask for which measures μ the Cauchy kernel defines a bounded operator on $L^2(d\mu)$. By the way, we are just talking about L^2 for convenience, but L^p, $1 < p < +\infty$, would do as well (and give the same measures μ), assuming μ is in the class $\sum$ of Definition 2.3 below.

We'll see on the way that the question is not so simple, even when $k = 1$ and $n = 2$. Let us at least justify our decision to restrict to measures $\mu \in \Delta$.

Proposition 1.4. *Let K_i, $i \in I$, be a finite family of good kernels such that, for some $\delta > 0$ and all $x \neq 0$, $| K_i(x) | > \delta | x |^{-k}$ for some $i \in I$. If μ is a non-atomic measure such that the T_μ^ϵ's corresponding to the various K_i's are uniformy bounded (in i and ϵ), then $\mu \in \Delta$.*

A minor variant of this result is proved in [Se3] allegedly using a well-known method; let us give here a slightly different proof.

Let us suppose that the T_μ^ϵ's are uniformly bounded, and that $\mu \notin \Delta$, and let us find an atom.

Lemma 1.5. *There is a constant C_0 such that if $Q_0 \subset \mathbb{R}^n$ is a cube, r_0 is its sidelength, $m_0 = \mu(Q_0)$ and $\lambda_0 = m_0 r_0^{-k}$, then one can find a cube $Q_1 \subset Q_0$, of radius $r_1 = r_0/100$, and such that $m_1 = \mu(Q_1) \geq \left(1 - \frac{C_0}{\lambda_0^2}\right) m_0$.*

Of course, this will be useful only when λ_0 is large enough ! Let us prove the lemma by contradiction. Split Q_0 into C_1^n cubes of size $C_1^{-1} r_0$; one of the cubes (call it R) is such that $\mu(R) \geq C_1^{-n} m_0$. Also, if the lemma is not true for Q_0, one can find another cube $S \subset Q_0$, of size $C_1^{-1} r_0$, and at distance $\geq \frac{r_0}{200}$ from R, with a measure $\mu(S) \geq \left[\frac{C_0}{\lambda_0^2} m_0\right] C_1^{-n}$.

Take the test-function $f = 1_S$, and look at the $T_\mu^\epsilon f(x)$, for $\epsilon << r$, on the cube R. If C_1 is large enough, then there is an $i \in I$ such that the corresponding $T_\mu^\epsilon f$ satisfy $| T_\mu^\epsilon f(x) | \geq \frac{1}{C}\mu(S) r_0^{-k}$ on R [this is because, the $K_i(x,y)$ being good kernels, each of them is almost constant when y varies in S]. So

$$\| T_\mu^\epsilon f \|_2 \geq \frac{1}{C}\mu(R)^{1/2}\mu(S) r_0^{-k} \geq \frac{1}{C}\mu(R)^{1/2}\mu(S)^{1/2} r_0^{-k} \| f \|_2,$$

and since the T_μ^ϵ's are uniformly bounded, we get $\mu(R)^{1/2}\mu(S)^{1/2} r_0^{-k} \leq C$. Using our estimates on $\mu(R)$ and $\mu(S)$, we get $m_0 C_1^{-n}\frac{C_0^{1/2}}{\lambda_0} \leq C r_0^k$ and, if C_0 is chosen large enough, $\frac{m_0}{\lambda_0} \leq \frac{1}{2} r_0^k$, which contradicts the definition of λ_0. So Lemma 1.5 is true.

Let us now use the lemma to show that, if λ_0 is large enough, there is a sequence $Q_0 \supset Q_1 \supset \cdots \supset Q_i \supset \cdots$ of cubes, with sidelengths $r_i = (100)^{-i} r_0$ and measures $m_i = \mu(Q_i)$ such that $\lambda_i = m_i r_i^{-k} > 10^i \lambda_0$.

We certainly have a cube Q_0 because $\mu \notin \Delta$. Suppose we constructed Q_i, and apply the lemma to Q_i. We find a cube $Q_{i+1} \subset Q_i$, with sidelength $r_{i+1} = r_i/100$ and mass $m_{i+1} \geq \left(1 - \frac{C_0}{\lambda_i^2}\right) m_i > \frac{1}{2} m_i$ if λ_0 is large enough. Then $\lambda_{i+1} = (100)^k \frac{m_{i+1}}{m_i}\lambda_i > 10\lambda_i$, as promised.

We get a better estimate for m_i than what was just written :

$$m_i \geq \left(1 - \frac{C_0}{\lambda_{i-1}^2}\right)\left(1 - \frac{C_0}{\lambda_0^2}\right) m_0 \geq \left(1 - \frac{C_0}{\lambda_0^2}(100)^{-i+1}\right)\cdots\left(1 - \frac{C_0}{\lambda_0^2}\right) m_0 \geq \frac{m_0}{2}$$

if we choose $\lambda_0^2 \geq 10 C_0$, for instance. Then $\bigcap_i \overline{Q_i}$, which is reduced to one point, has a measure $\geq \frac{m_0}{2} > 0$, and the proposition is true.

Remark 1.6. If $\mu \in \Delta$ and K is a good kernel, one can define $T_\mu^\epsilon f(x)$ everywhere for any $f \in L^2(d\mu)$ (use Schwarz' inequality, and the estimate

$$\int_{|x-y|>\epsilon} | x - y |^{-2k} \, d\mu(y) =$$

$$\sum_{\ell=0}^{\infty} \int_{|x-y|\sim 2^\ell \epsilon} | x - y |^{-2k} \, d\mu(y) \leq \sum_{\ell=0}^{\infty} C(2^\ell\epsilon)^{-2k}(2^\ell\epsilon)^k < +\infty).$$

Example 1.7. Let $z : \mathbb{R}^k \to \mathbb{R}^n$ be a continuous function. We think of z as being the parametrization of some surface. We can define a measure μ on $\mathbb{R}^n$ by $\mu(f) = \int_{\mathbb{R}^n} f \, d\mu = \int_{\mathbb{R}^k} f(z(x))dx$ (for $f \in C_c(\mathbb{R}^n)$). We'll often call μ the "measure associated with z".

Saying that $\mu \in \Delta$ amounts to saying that

(6) $$\left|\left\{x \in \mathbb{R}^k \ : \ z(x) \in B(w,r)\right\}\right| \leq C r^k$$

for all $w \in \mathbb{R}^n$ and all $r > 0$.

2. Calderón-Zygmund techniques

We shall need extensions to our context, of some of the classical results of "Calderón-Zygmund theory". By lack of time, we'll only recall the main steps of the proof ; if the reader has problems filling the gaps, he will find all the details in [Dv1], [Dv2], or [Dv5].

One of the useful tricks that will be used here is distinguishing between the measure μ that comes into the definition of T^*_μ, and the measure (call it σ) on which we want to integrate $T^*_\mu(x)$. This is a little like studying a bilinear form instead of a quadratic form. The advantage is that we'll be able to change μ (or σ) without affecting the other measure.

We need an easy extension of the usual maximal theorem concerning the Hardy-Littlewood maximal function.

Definition 2.1. If μ is a nonnegative Radon measure and f is measurable, define a maximal function by

$$M_\mu f(z) = \sup_{r>0} r^{-k} \int_{|z-w|\leq r} |f(w)| \, d\mu(w).$$

Note that $M_\mu f$ is allowed to be $+\infty$ at some points ; when $k = n$, $M_\mu f$ is the usual Hardy-Littlewood maximal function of $f \, d\mu$.

Lemma 2.2. *If μ and σ are two measures in Δ, $1 < p < +\infty$ and $f \in L^p(d\mu)$, then $M_\mu f \in L^p(d\sigma)$, with*

$$\| M_\mu f \|_{L^p(d\sigma)} \leq C(p,\mu,\sigma) \| f \|_{L^p(d\mu)} . \tag{7}$$

The proof is the same as usual : one uses interpolation to reduce to showing that M_μ sends $L^\infty(d\mu)$ into $L^\infty(d\sigma)$ (which is trivial), and also $L^1(d\mu)$ into weak-$L^1(d\sigma)$, which can be proved easily with the help of the Besicovitch covering lemma.

Definition 2.3. We shall denote by $\sum$ the subset of Δ composed of the measures σ that also satisfy, for some $\gamma > 0$,

$$\sigma(B(w,r)) \geq \gamma r^k \text{ for all } w \in \operatorname{supp}\sigma \text{ and } r > 0. \tag{8}$$

Example 2.4. If $z : \mathbb{R}^k \to \mathbb{R}^n$ is such that the measure μ associated to z is in Δ, and if z is Lipschitz (i.e., $| z(x) - z(y) | \leq C | x - y |$), then $\mu \in \sum$.

Proof. If $w \in \operatorname{supp}\mu$ and x is such that $z(x) = w$, then $\mu(B(w,r)) = \mid z^{-1}(B(w,r)) \mid \geq \mid B(x, \frac{r}{C}) \mid \geq C' r^k$.

In particular, if μ is the arclength measure on a connected, rectifiable curve of infinite length, the condition (8) is always satisfied.

The following lemma can be seen as an analogue of the result on quadratic forms that says that a bilinear form is dominated by the corresponding quadratic form.

Lemma 2.5. *Let K be a good kernel, $\mu \in \Delta$ and $\sigma \in \sum$. Suppose that, for all $1 < p < +\infty$, T_σ^* is bounded on $L^p(d\sigma)$. Then, for $1 < p < +\infty$, T_σ^* is bounded from $L^p(d\sigma)$ to $L^p(d\mu)$, and T_μ^* is bounded from $L^p(d\mu)$ to $L^p(d\sigma)$.*

The proof of Lemma 2.5 relies on pointwise inequalities. For instance, the boundedness of T_σ^* from $L^p(d\sigma)$ to $L^p(d\mu)$ follows from the inequality

$$T_\sigma^* f(z) \leq C \left[M_\sigma(T_\sigma^* f) \right](z) + C M_\sigma f(z) \tag{9}$$

for all $z \in \mathbb{R}^n$, and from Lemma 2.2.

To prove that T_μ^* sends $L^p(d\mu)$ into $L^p(d\sigma)$, one first proves that there is an operator T_μ, which is obtained as a weak limit of T_μ^ϵ's, and which is bounded from $L^p(d\mu)$ to $L^p(d\sigma)$. The existence of T_μ follows from the fact that the T_σ^ϵ's are uniformly bounded from $L^p(d\sigma)$ to $L^p(d\mu)$ (which follows from the boundedness of T_σ^* on the same spaces), and duality. To deduce the boundedness of T_μ^* (from $L^p(d\mu)$ to $L^p(d\sigma)$) from the existence of T_μ, one uses the following generalization of Cotlar's inequality :

$$T_\mu^* f(z) \leq C \left[M_\sigma(T_\mu f) \right](z) + C \left[M_\mu \mid f \mid^{\sqrt{p}} \right]^{1/\sqrt{p}} (z). \tag{10}$$

The boundedness of T_μ^* easily follows from (10) and Lemma 2.2.

The proof of the pointwise inequalities (9) and (10) is relatively straightforward. The only things that are used are the estimates on K and its first derivative. The proofs are given in full details in [Dv1] in dimension 1 (but the argument is the same in all dimensions). One can also look at [Dv2] or [Dv5].

Remark. In the statement of Lemma 2.5, we assumed that T_σ^* is bounded on all $L^p(d\sigma)$'s. This is only for convenience, because one can prove that the L^p-boundedness for one p implies the L^p-boundedness for all other p, $1 < p < +\infty$. The proof is a routine Calderón-Zygmund argument, using the fact that if $\sigma \in \sum$, the support of σ is a space of homogeneous type.

The machine which will be described in the next section will need an initial result to get started. We shall feed it with the following special case.

Lemma 2.6. *If K is a good kernel and σ is the surface measure on the graph of a Lipschitz function $A : \mathbb{R}^k \to \mathbb{R}^{n-k}$, then the operator T_σ^* is bounded on $L^p(d\sigma)$ for $1 < p < +\infty$.*

(Of course, the image by a (linear) isometry of such a graph would work as well !)

When the kernel K is antisymmetric and homogeneous of degree k, this lemma is a rather easy consequence of Coifman, McIntosh and Meyer's theorem (more precisely, of Example 8.5 of Part II) and the rotation method. When K is a general "good kernel" (as in Definition 1.1), Lemma 2.6 follows easily from Example 6.7 of Part II : the only difference is that we replaced the Lebesgue measure on $\mathbb{R}^k$ by the surface measure on the graph, and also that we used a slightly different truncation from the usual one (the difference is easily estimated by a maximal function).

3. The "good λ" method.

Good λ inequalities are an impressive tool, invented some time ago by D. Burkholder and R. Gundy. Here is a version of their lemma.

Lemma 3.1 *Let X be a set and μ a measure on X. Let $u : X \to [0, +\infty]$ be measurable. Suppose u is equal, except perhaps on a set of finite measure, to a function of $L^p(X, d\mu)$. Let $v \in L^p(X, d\mu)$. Suppose that there is a $0 < \eta < 1$ and, for each $\epsilon > 0$, a constant $\gamma > 0$ such that, for all $\lambda > 0$,*

(11) $\mu(\{x \in X : u(x) > \lambda + \epsilon\lambda \text{ and } v(x) \leq \gamma\lambda\}) \leq (1-\eta)\mu(\{x \in X : u(x) > \lambda\}).$

Then $u \in L^p(X, d\mu)$ and $\| u \|_p \leq C(p, \epsilon, \eta, \gamma) \| v \|_p$.

We'll leave the proof as an exercise (one can also go and look at [Dv1] or [Dv2]). We do not need every ϵ, but the ϵ's that work depend on η and p ; it will be just as easy to find a γ for each $\epsilon > 0$.

The hypothesis of the lemma is some reasonably subtle way of saying that u is dominated by v. The surprising fact is how well Lemma 3.1 works. The following result might seem quite technical. We shall see later how to use it (see Definition 3.4 and Corollary 3.6 below).

Proposition 3.2. *Let $\mu \in \sum$; suppose that there exist constants $0 < \theta < 1$, $C \geq 0$ and, for each $1 < p < +\infty$, $A_p \geq 0$ such that the following is true. For each ball B centered on the support of μ, one can find a measure $\sigma \in \sum$ and a compact set $E \subset B \cap \operatorname{supp}\mu$ such that*

(12) *σ satisfies (3) with the constant C and (8) with the constant $\gamma = \frac{1}{C}$;*

(13) $\mu(E) \geq \theta\mu(B)$;

(14) $1_E\mu \leq \sigma$ *(i.e.: σ is larger than the restriction of μ to E), and*

(15) $\| T_\sigma^* f \|_{L^p(d\sigma)} \leq A_p \| f \|_{L^p(d\sigma)}$ for $f \in C_c(\mathbb{R}^n)$.

Then, there are constants C_p, $1 < p < +\infty$, such that

(16) $\| T_\mu^* f \|_{L^p(d\mu)} \leq C_p \| f \|_{L^p(d\mu)}$ for $f \in C_c(\mathbb{R}^n)$.

The proof is quite standard, but we give it to convince the unexperienced reader that good λ inequalities are a really powerful tool. Note that the proposition says, in some sense, that local information on a small piece of the support of μ, at all scales, is enough to obtain a global result of boundedness on $L^2(d\mu)$. This is a little similar to John and Nirenberg's theorem on BMO.

Proof. Given $f \in C_c(\mathbb{R}^n)$, we want to apply Lemma 3.1 with $u(x) = T_\mu^* f(x)$. Let us start by checking the qualitative assumption.

Lemma 3.3. *The function u is equal, outside of a compact, to a function of $L^p(d\mu)$.*

Indeed, if $|z|$ is large enough, $|T_\mu^* f(z)| \leq \int_{\operatorname{supp} f} |K(z-w)| |f(w)| d\mu(w) \leq C M_\mu f(z)$, which belongs to $L^p(d\mu)$ by Lemma 2.2.

Let us now choose the function v :

$$v(x) = M_\mu f(x) + \{M_\mu(|f|^r)\}^{1/r}(x), \tag{17}$$

where $r = \sqrt{p}$, for instance. We want to prove (11).

Let $\lambda > 0$, and call $\Omega = \{x \in \mathbb{R}^n : u(x) > \lambda\}$; Ω is open because T_μ^* is lower semi-continuous. Cover $\Omega \cap \operatorname{supp} \mu$ by the balls $B(x) = B\left(x, \frac{1}{2}\operatorname{dist}(x, \Omega^c)\right)$, where $x \in \operatorname{supp} \mu$. A covering lemma of Vitali type (see [St], p. 9-10) gives a sequence of points x_i such that the $B(x_i)$ are disjoint, but $\Omega \cap \operatorname{supp} \mu \subset \bigcup_{i=1}^\infty 10B(x_i)$.

If we prove that, for each i,

$$\mu(\{x \in B(x_i) \ : \ u(x) \leq \lambda + \epsilon\lambda \ \text{ or } \ v(x) > \gamma\lambda\}) \geq \frac{\theta}{2}\mu(B(x_i)), \tag{18}$$

then (11) will follow because summing on i gives

$$\mu(\{x \in \Omega \ : \ u(x) \leq \lambda + \epsilon\lambda \text{ or } v(x) > \gamma\lambda\}) \geq \frac{\theta}{2}\sum_i \mu(B(x_i)) \geq \frac{\theta}{C}\sum_i \mu(10B(x_i))$$

(because $B(x_i)$ is centered on $\operatorname{supp}\mu$ and $\mu \in \sum) \geq \frac{\theta}{C}\mu(\Omega)$. This is (11), with $\eta = \frac{\theta}{C}$.

Thus, we only need to find, for each $\epsilon > 0$, a $\gamma > 0$ such that (18) is satisfied for each i. If $\nu(x) > \gamma\lambda$ for all $x \in B(x_i)$, there is nothing to prove. So we can suppose that there is a $\xi \in B(x_i)$ such that $v(\xi) \leq \gamma\lambda$.

Write $f = f_1 + f_2$, with $f_1 = 1_{B(\xi,R)} f$, and where $R = C_1 \operatorname{dist}(\xi, \Omega^c)$ for some large constant C_1. For f_2, one uses the regularity of K to show that, if C_1 is large enough, $T_\mu^* f_2(x) \leq T_\mu^* f(a) + C M_\mu f(\xi)$ for all $x \in B(x_i)$ and $a \in 3B(x_i)$. (The verification is left as an exercise.) Let us choose a point $a \in 3B(x_i) \cap \Omega^c$, so that we get

$$T_\mu^* f_2(x) \leq \lambda + \frac{\epsilon\lambda}{2} \quad \text{for all } \ x \in B(x_i), \tag{19}$$

provided we choose γ small enough.

Therefore, we only need to prove that

$$(20) \qquad \mu(\{x \in B(x_i) \ : \ T_\mu^* f_1(x) \leq \frac{\epsilon\lambda}{2}\}) \geq \frac{\theta}{2}\mu(B(x_i)).$$

Let us use the compact set E and the measure σ of the hypothesis. Since $\mu(E) \geq \theta\mu(B(x_i))$, we only have to show that $\mu(\{x \in E \ : \ T_\mu^* f_1(x) > \frac{\epsilon\lambda}{2}\}) \leq \frac{\theta}{2}\mu(B(x_i))$. By (14), the left hand side is less than $\sigma(\{x \in E \ : \ T_\mu^* f_1(x) > \frac{\epsilon\lambda}{2}\}) \leq 2^r \epsilon^{-r}\lambda^{-r} \parallel T_\mu^* f_1 \parallel_{L^r(d\sigma)}^r$.

By (16), T_σ^* is bounded on $L^p(d\sigma)$ for $1 < p < +\infty$. By Lemma 2.5, T_μ^* sends $L^r(d\mu)$ to $L^r(d\sigma)$, and so $\mu(\{x \in E \ : \ T_\mu^* f_1(x) > \frac{\epsilon\lambda}{2}\}) \leq C\epsilon^{-r}\lambda^{-r} \parallel f_1 \parallel_{L^r(d\mu)}^r \leq C\epsilon^{-r}\lambda^{-r}R^k\{M_\mu(| f |^r)\}(\xi) \leq C\epsilon^{-r}\lambda^{-r}\mu(B(x_i))\gamma^r\lambda^r \leq \frac{\theta}{2}\mu(B(x_i))$ if we choose γ small enough.

This proves (20), and so (11) and the proposition follow.

Let us say how we intend to use Proposition 3.2. Let $S \subset \mathbb{R}^n$ be given with a measure $\mu \in \sum$ such that $S = \operatorname{supp} \mu$.

Definition 3.4. We shall say that S "contains big pieces of Lipschitz graphs" (and often write S CBPLG) if there are constants $0 < \theta < 1$ and $M \geq 0$ such that, for each $x \in S$ and $r > 0$, there is a compact set $E \subset S \cap B(x,r)$, with $\mu(E) \geq \theta r^k$, and which is contained in the image by a linear isometry of $\mathbb{R}^n$ of the graph of some M-lipschitz function $A \ : \ \mathbb{R}^k \to \mathbb{R}^{n-k}$.

Remark 3.5. The notion is really a notion concerning S, and not μ, because if $S \subset \mathbb{R}^n$ is the support of some measure $\mu \in \sum$, then μ is equivalent to the k-dimensional Hausdorff measure on S (the verification is a standard covering argument, which is left as an exercise). Of course, the property of Definition 3.4 does not change when μ is replaced by any measure σ such that $C^{-1}\mu \leq \sigma \leq C\mu$!

Corollary 3.6. *If S CBPLG, and $\mu \in \sum$ is such that $\operatorname{supp} \mu = S$, then T_μ^* is bounded on $L^p(S, d\mu)$ for $1 < p < +\infty$.*

To prove the corollary, one applies Proposition 3.2, with for σ the k-dimensional surface measure on the image by an isometry of the Lipschitz graph of the definition. The estimate (15) then comes from Lemma 2.6, and (14) follows because both $1_E\mu$ and $1_E\sigma$ are equivalent to the k-dimensional Hausdorff measure on E (see the remark above, or the argument in [Dv5], p. 252).

Exercise. Show that if $z \ : \ \mathbb{R}^k \to \mathbb{R}^n$ is a bilipschitz mapping (i.e. satisfies $C^{-1} \mid u - v \mid \leq \mid z(u) - z(v) \mid \leq C \mid u - v \mid$ for some $C > 0$), then $S = z(\mathbb{R}^k)$ CBPLG (and so Corollary 3.6 applies to S). [Warning : this is more challenging than most of the previous exercises. If you don't find, just go on reading !]. This is not the most striking application of Corollary 3.6, because the boundedness of the singular integral operators on S can be proved directly, using the parametrization of S and the rotation method, as for Lemma 2.6 or Example 8.5 of Part II.

Remark 3.7. One can imagine a more complicated corollary. Let us say that " $S(\mathrm{CBP})^2\mathrm{LG}$" if S "contains big pieces of surfaces that CBPLG uniformly" (with a definition similar to Definition 3.4). A second use of Proposition 3.2 gives, exactly like in the proof of Corollary 3.6, that T_{μ}^{*} is bounded on all $L^p(S)$. One can of course go further, and prove the boundedness of T_μ^* whenever $S(\mathrm{CBP})^m\mathrm{LG}$ for some m (with obvious definitions).

We shall come back to this in later sections.

We conclude this section by a remark on the existence of principal values. Up to now, we tried to avoid talking about "the operator defined by the kernel K", because the integral $\int K(x-y)f(y)d\mu(y)$ does not converge in general. It turns out that, once we know that T_μ^* is bounded, very little is needed to conclude that T_μ^ϵ converges to an operator T_μ as ϵ tends to 0.

Proposition 3.8. *Suppose k is an integer, K is a good kernel and $\mu \in \sum$ is such that T_μ^* is bounded on $L^2(d\mu)$. Also suppose that the support S of μ is a rectifiable set. Then, for all $f \in L^2(d\mu)$, $T_\mu f(x) = \lim_{\epsilon \to 0} T_\mu^\epsilon f(x)$ exists for μ-almost every x.*

There are various (non-trivially) equivalent definitions of a rectifiable set, and we'll choose the one that makes the proposition easy. We'll say that S is rectifiable if it is contained in the union of a set of measure 0 and a countable number of (rotated) Lipschitz graphs. For more information about rectifiable sets, see [Fe], [Mt] or [Fa].

Exercise. Show that if S CBPLG, or even if $S(\mathrm{CBP})^{\mathrm{m}}\mathrm{LG}$ (see Remark 3.7), then S is rectifiable.

We stated an L^2-result here, but the corresponding L^p-result also holds, even up to $p = 1$ included. The proof of Proposition 3.8 (or of its L^p-variant) only requires fairly standard techniques, so we'll leave it as an exercise, with the following instructions.

1. First reduce to the case when μ is the restriction to S of the k-dimensional Hausdorff measure (remember Remark 3.5).

2. Next, note that it is enough to prove that $\lim_{\epsilon \to 0} T_\mu^\epsilon f(x)$ a. e. for a dense class of functions f. Indeed, given $f \in L^2(d\mu)$, it is enough to show that

$$\mathcal{O} f(x) = \mid \limsup_{\epsilon, \epsilon' \to 0} T_\mu^\epsilon f(x) - T_\mu^{\epsilon'} f(x) \mid$$

has an L^2-norm which is equal to 0. This is shown by writing $\mathcal{O} f \leq \mathcal{O} g + 2T_\mu^*(f-g)$, and by choosing g in the dense class and close enough to f. This argument is a fairly standard way to use maximal estimates (see [St], p. 8 or 45, for instance).

3. Now suppose the result was established when S is a Lipschitz graph. Take for your dense class the linear combinations of functions supported on the intersection of S with Lipschitz graphs (here we use the rectifiability of S). Since $\lim_{\epsilon \to 0} T_\mu^\epsilon f(x)$ always exists when $x \notin \overline{\text{supp } f}$, the general case will follow.

4. Prove the proposition when $k = 1$ and S is a straight line. For this case, compactly supported smooth functions are a nice enough dense class.

5. Show that, when k is odd, the operator of Example 6.7 of Part II is a principal value operator in the classical sense : for $f \in L^2$, $\lim_{\epsilon \to 0} \int_{|u-v|>\epsilon} K(u,v) f(v) dv$ exists for a.e. u. To do this, first note that each time the boundedness of an operator is established, the corresponding maximal operator is also bounded, because of Cotlar's inequality (or (III.10)). Therefore, one only has to follow the proof and apply the dominated convergence theorem as often as needed.

6. Conclude. You may have to use the fact that the graph of a Lipschitz function has tangent planes almost everywhere to go from the operator of Example II.6.7 to the operator of Lemma 2.6.

4. Regular curves and Lipschitz graphs

Let us pause in a moment to see how the "good λ method" works when $k = 1$. Let us first consider the case of a connected, rectifiable curve Γ ; for simplicity, we shall restrict to the case when Γ has infinite length.

A - Regular curves.

Definition 4.1. A (connected, rectifiable) curve Γ is regular if there is a constant $C \geq 0$ such that, for all $x \in \mathbb{R}^n$ and $r > 0$, the total length of $\Gamma \cap B(x,r)$ is less than Cr.

The notion is due to Ahlfors [Ah]. If μ is the arclength measure on Γ, we see that Γ is regular if and only if $\mu \in \Delta$. Because Γ is connected, μ also satisfies (8) automatically, and so $\mu \in \sum$.

Example. Chord-arc curves are regular ; a parabola in $\mathbb{R}^2$ is regular (but not chord-arc).

Theorem 4.2. *Let $\Gamma \subset \mathbb{R}^n$ be a regular curve, and μ the arclength measure on Γ. Then T^*_μ (as defined by (4) and (5)) is bounded on $L^p(\Gamma, d\mu)$ for $1 < p < +\infty$, when K is any "good kernel" (see Definition 1.1, with $k = 1$).*

Note that Proposition 1.4 says that the condition is also necessary. To prove the theorem, we shall appeal to Corollary 3.6, and show that Γ CBPLG.

Let $x \in \Gamma$ and $r > 0$. Call $z(s)$, $0 \leq s \leq s_0$ a parametrization by arclength of the piece of Γ between x and the first time Γ gets out of $B(x,r)$. Thus, $z(0) = x$ and $| z(s_0) - x | = r$. Since Γ is regular, we get $s_0 \leq C_0 r$ for some $C_0 > 0$. We shall find our big piece of Lipschitz graph in $z([0, s_0])$.

After a possible change of coordinates, we can assume that $x = 0$ and $z(s_0) = (r, 0, \cdots 0)$. Let $\delta = \frac{1}{2C_0}$.

Lemma 4.3. *There are a set $E \subset [0, s_0]$ and a Lipschitz function $h : [0, s_0] \to \mathbb{R}$, such that $h(s) = z_1(s)$ on E ($z_1(s)$ is the first coordinate of $z(s)$),*

(21) $$|E| \geq \frac{r}{2}$$

and

(22) $$\delta \leq h' \leq 1.$$

If we can prove this lemma, we will be finished for the following reason : s will be a Lipschitz function of $h(s)$ on $[0, h(s_0)]$, and so all the coordinates of $\tilde{z}(s) = (h(s), z_2(s), \cdots z_n(s))$ will be Lipschitz functions of the first coordinate, so that $\tilde{z}$ will be the parametrization of a Lipschitz graph. Then $z(E) = \tilde{z}(E)$ will be the big piece we are looking for, because $|z(E)| = |\tilde{z}(E)| \geq \delta |E| \geq \frac{\delta r}{2}$.

To prove Lemma 4.3, let us try the function $h(s) = \sup_{0 \leq t \leq s} [z_1(t) - \delta t] + \delta s$.

Certainly, h is Lipschitz and $h'(s) \geq \delta$. We have to check that if $E = \{s \in [0, s_0] : z_1(s) = h(s)\}$, then $|E| \geq \frac{r}{2}$. Note that

$$E = \left\{ s \in [0, s_0] \ : \ \sup_{0 \leq t \leq s} (z_1(t) - \delta t) = z_1(s) - \delta s \right\}.$$

As the reader might have guessed, we are about to apply the rising sun lemma. Let $f(s) = z_1(s) - \delta s$ and $F(s) = \sup_{0 \leq t \leq s} f(t)$ (see figure 1).

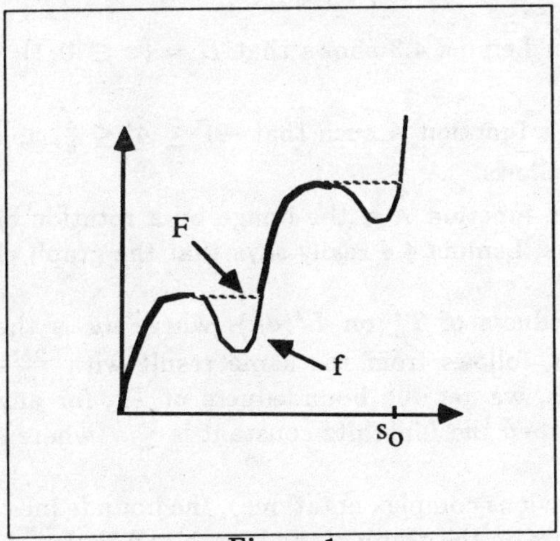

Figure 1

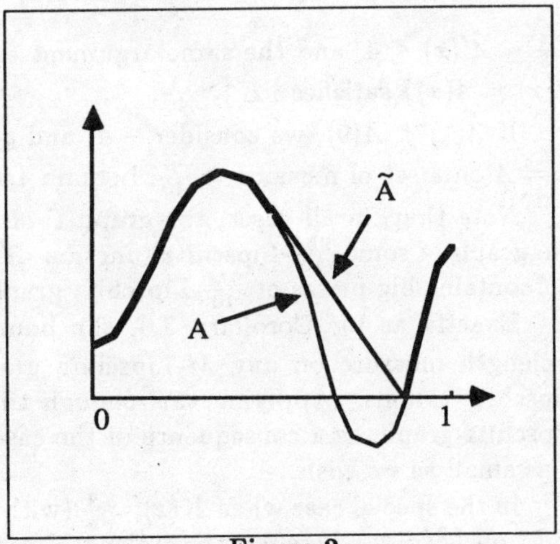

Figure 2

The set $\Omega = [0, s_0] \setminus E$ is open, and so $\Omega = \bigcup_k I_k$, where the I_k's are disjoint open intervals. Calling $I_k =]a_k, b_k[$, we have $f(a_k) \geq f(b_k)$ (with equality, except perhaps on the last component). Then $f(s_0) - f(0) = r - \delta s_0 \geq r/2$, and so $\frac{r}{2} \leq \int_0^{s_0} f'(t)dt = \int_E f'(t)dt + \sum_k \int_{a_k}^{b_k} f'(t)dt \leq | E | + \sum_k (f(b_k) - f(a_k)) \leq | E |$, which is (21).

This proves Theorem 4.2.

B - The Cauchy integral on Lipschitz graphs again

Proposition 3.2 can also be used to derive Coifman, McIntosh and Meyer's theorem from Calderón's. The idea will be to show that M-Lipschitz graphs (i.e., graphs of M-Lipschitz functions) "contain big pieces of $\frac{9M}{10}$-Lipschitz graphs" (with a definition similar to Definition 3.4).

Lemma 4.4. *Let $A : \mathbb{R} \to \mathbb{R}$ be a Lipschitz function, and let $M = \| A' \|_\infty$. For each compact interval $I \subset \mathbb{R}$, there is a Lipschitz function $\tilde{A}$ and a compact $E \subset I$ such that $\tilde{A}(x) = A(x)$ for $x \in E$,*

$$| E | \geq | I | /10, \tag{23}$$

and

$$-M \leq \tilde{A}'(x) \leq \frac{4M}{5} \quad a.\ e.,\ or\ else \quad -\frac{4M}{5} \leq \tilde{A}'(x) \leq M \quad a.\ e. \tag{24}$$

To prove the lemma, one can easily reduce to the case when $M = 1$, and also $I = [0, 1]$. First suppose that $A(1) \geq A(0)$. Pick $\tilde{A}(x) = \sup_{0 \leq t \leq x} [A(t) + \frac{4}{5}t] - \frac{4}{5}x$ (see Figure 2) ; then $-\frac{4}{5} \leq \tilde{A}'(x) \leq 1$, and the same argument as in Lemma 4.3 shows that $E = \{x \in [0, 1] : \tilde{A}(x) = A(x)\}$ satisfies $| E | \geq \frac{1}{10}$.

If $A(1) < A(0)$, we consider $-A$, and get a function $\tilde{A}$ such that $-1 \leq \tilde{A}' \leq \frac{4}{5}$, and $\tilde{A} = A$ on a set of measure $\geq \frac{1}{10}$. Lemma 4.4 follows.

Note that, in all cases, the graph $\tilde{\Gamma}$ of the function $\tilde{A}$ is the image by a rotation of the graph of some $\frac{9M}{10}$-Lipschitz function. Thus, Lemma 4.4 really says that the graph of A "contains big pieces of $\frac{9M}{10}$-Lipschitz graphs".

Exactly as for Corollary 3.6, the boundedness of T_μ^* on $L^2(d\mu)$, where $d\mu$ is the arclength measure on any M-Lipschitz graph, follows from the same result with $\frac{9M}{10}$-Lipschitz graphs. Applying this enough times, we get the boundedness of T_μ^*, for any Lipschitz graph, as a consequence of the case when the Lipschitz constant is $\leq \epsilon$ (where ϵ is as small as we wish).

In the special case when $K(z) = \frac{1}{z}$ (with obvious complex notations), the boundedness of T_μ^* on $L^2(d\mu)$, where μ is the arclength measure on the graph of the Lipschitz function A, is equivalent to the boundedness of the operator C_A of Example II.6.3. Thus we deduced Coifman-McIntosh-Meyer's result from Calderón's (C_A is bounded if $\| A' \|_\infty$ is small enough).

Exercise. Check that if we know that $\| T_\mu^* \| \leq \lambda$ for $\frac{9M}{10}$-Lipschitz graph, the proof gives $\| T_\mu^* \| \leq C(\lambda + M + 1)$ for M-Lipschitz graphs. Thus, we get the estimate $\| T_\mu^* \| \leq C(1+M)^N$ for some N. You can now complete the proof of Example 6.7 of Part II. Of course, we did not use Lemma 2.6 in part B of this section, and so the reader will agree that we did not introduce any vicious circle.

Remark. The argument above seems quite crude, and also quite complicated, because we used the machinery of Sections 2 and 3. However, in the special case of Lipschitz curves, the argument can be simplified a great deal. What is more surprising is that one can get very good estimates with an argument like that ! After partial results by Murai and Tchamitchian, Murai proved the best estimate for C_A : $\| C_A \| \leq C(1+ \| A' \|_\infty)^{1/2}$, using a (quite refined !) argument of the same type (see [Mu1] or [Mu2]). With a similar proof, he also shows that the kernel $\frac{1}{x-y} \exp i \left[\frac{A(x)-A(y)}{x-y}\right]$ defines a bounded operator on $L^2(\mathbb{R})$, with norm $\leq C(1+ \| A' \|_\infty)$.

5. Garnett's example

The following example was introduced by Ivanov (in his thesis), and J. Garnett [Ga1] to study (related) problems of analytic capacity.

It is the Cantor set $H = k \times k \subset \mathbb{R}^2 \simeq \mathbb{C}$, where k is the "middle half" Cantor set on the line. More precisely, $k = \bigcap_n k_n$, where $k_0 = [0,1]$, $k_1 = [0, \frac{1}{4}] \cup [\frac{3}{4}, 1]$, k_n is composed of 2^n intervals of length 4^{-n}, and k_{n+1} is obtained from k_n by removing, from each interval of k_n, the open interval of length $\frac{1}{2}4^{-n}$ situated in its center (see a picture of $k_2 \times k_2$ below).

The constant $\frac{1}{4}$ is chosen so that H is a one-dimensional set. There is a natural measure on H : it is the measure μ, supported on H, such that, for each n, the measure of each of the 4^n squares of sidelength 4^{-n} composing $k_n \times k_n$ is 4^{-n} (μ is proportional to the 1-dimensional Hausdorff measure on H). Note that $\mu \in \sum$ (only locally, since (8) is not satisfied for large r's ; one could easily modify the example to get a $\mu \in \sum$).

The main interest for us of the set H is that, although $\mu \in \sum$, the Cauchy integral does not define a bounded operator on $L^2(d\mu)$. One can prove this by showing that H has zero analytic capacity (see [Ga1]) and then using duality. Let us very rapidly sketch the idea of a direct proof (the computations can be found in [Dv3]).

Let T_n be defined by

$$(25) \qquad T_n f(z) = \int_{\frac{1}{2}4^{-n} \leq |z-w| \leq \sqrt{2}4^{-n}} (z-w)^{-1} f(w) d\mu(w)$$

(if z is in H, this consists in integrating on the squares of $k_n \times k_n$ that do not contain z).

The idea is that the functions $f_n = T_n 1$ have the same L^2-norm, and behave a little as if they were independent (because they vary at different scales and have integral zero). The proof then consists in estimating the scalar products $< f_m, f_n >$, and showing that they vanish fast enough, so that one gets

$$\| \sum_{0 \le n \le n_0} T_n 1 \|_2^2 \ge \frac{n_0}{C}. \tag{26}$$

There is also a much less computational proof, by P. Jones, of the fact that the Cauchy kernel does not define a bounded operator on $L^2(d\mu)$, which has the advantage of being much more flexible (see [Jn1]). Unfortunately, I don't think it gives the estimate (26).

Remark 5.1. It is not too hard to transform (26) into an estimate for the Cauchy integral on Lipschitz graphs. Notice that, if we change coordinates (or rotate H), the set H (minus a countable set of corners) becomes the graph of a function (see the picture).

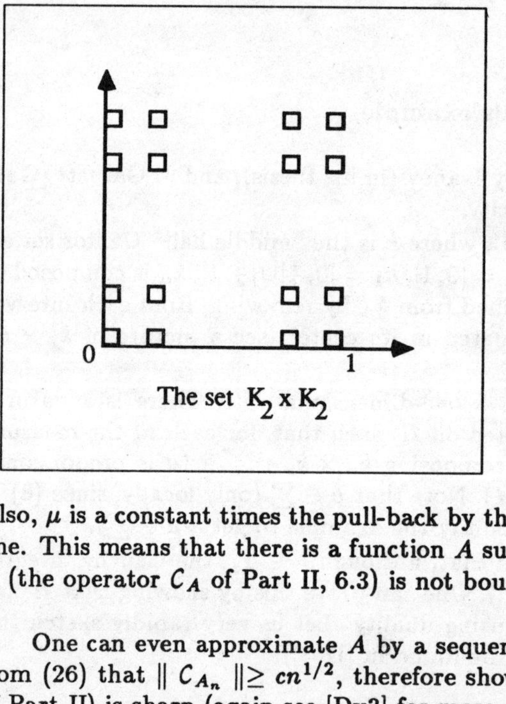

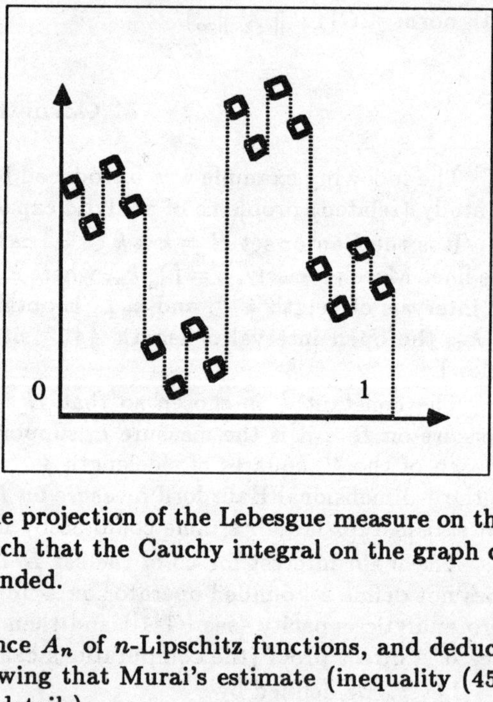

Also, μ is a constant times the pull-back by the projection of the Lebesgue measure on the line. This means that there is a function A such that the Cauchy integral on the graph of A (the operator C_A of Part II, 6.3) is not bounded.

One can even approximate A by a sequence A_n of n-Lipschitz functions, and deduce from (26) that $\| C_{A_n} \| \ge cn^{1/2}$, therefore showing that Murai's estimate (inequality (45) of Part II) is sharp (again see [Dv3] for more details).

Remark 5.2. The fact that the Cauchy integral is not bounded on $L^2(d\mu)$ shows that the connectedness assumption of Definition 4.1 and Theorem 4.2 was not superfluous. If Γ_n is the boundary of $k_n \times k_n$, and μ_n is the arclength measure on Γ_n, we see that $\mu_n \in \sum$ uniformly in n ; however, since $\mu_n \to \mu$, the $T^*_{\mu_n}$'s are not uniformly bounded !

One can go a little further, and build a counterexamaple in higher dimensions. For instance, let S_n be the 2-dimensional surface of $\mathbb{R}^3$ defined by $S_n = k_n \times k_n \times \{1\} \cup$

$\Gamma_n \times [0,1] \cup [\mathbb{R}^2 \backslash (k_n \times k_n)] \times \{0\}$ (see the picture).

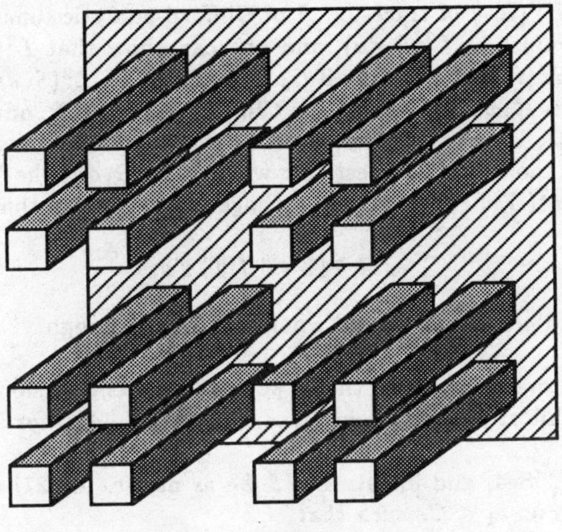

A sketch of the surface S_2

Note that the surface measure on S_n is a measure μ_n which is in $\sum$, uniformly in n, but it is not hard to find a kernel K such that the corresponding $T^*_{\mu_n}$'s are not uniformly bounded on $L^2(\mu_n)$. Thus, in higher dimensions, something stronger than connectedness is required (see Sections 9 and 10 for further comments).

6. Three classes of surfaces

We wish to describe, rather rapidly, three classes of K-dimensional objects in $\mathbb{R}^n$ where something positive can be said about the operators T^*_μ's.

A - Chord-arc surfaces with small constant

This first example was introduced by S. Semmes ([Se3] and [Se4]). For this example, $k = n - 1$, and also one is interested in more refined properties than the boundedness of the T^*_μ's. We shall try to say, as fast as possible, what the point is (and leave all the definitions, precise statements, and proofs).

Let S be a hypersurface in $\mathbb{R}^{k+1}$. We shall make the a priori assumption that S is smooth (including at ∞), oriented, and separates $\mathbb{R}^{k+1}$ into two connected components (call them Ω^+ and Ω^-). Let $d\mu$ denote the surface measure on S.

One can define on S the analogue of the Cauchy operator. This time, the kernel takes values in the Clifford algebra (instead of $\mathbb{C}$ when $k = 1$). Let us call C_S the resulting operator (see Semmes' papers for definitions).

One of the nice things about the Clifford algebra setting is that there is a notion (we'll call it "Clifford-analyticity") that generalizes analyticity. One can then define Hardy spaces $H^2(\Omega^+)$ and $(H^2(\Omega^-)$ of traces on S of Clifford-analytic functions in Ω^+ and Ω^- that have boundary values in $L^2(S, d\mu)$, and one can show that $L^2(S, d\mu)$ is the direct sum of $H^2(\Omega^+)$ and $H^2(\Omega^-)$ if and only if C_S is bounded on $L^2(S, d\mu)$.

The kernel of C_S is $K(x-y)$, where K is Clifford valued, C^∞, odd, and homogeneous of degree $-k$, and so C_S is one of the operators of the previous sections.

S. Semmes asks a more precise question : when is $L^2(S, d\mu)$ the "almost-orthogonal" sum of $H^2(\Omega^+)$ and $H^2(\Omega^+)$? By "almost orthogonal", we mean that

$$|< f, g >| \leq \epsilon \parallel f \parallel \parallel g \parallel \tag{27}$$

for all $f \in H^2(\Omega^+)$ and $g \in H^2(\Omega^-)$, and where ϵ is small enough.

In terms of operators, this means that a slightly different version $\tilde{C}_S$ of the Clifford-Cauchy operator is "almost antiselfadjoint" (i.e. $\parallel \tilde{C}_S + \tilde{C}_S^* \parallel \leq \epsilon'$ for an $\epsilon' \sim \epsilon$).

Theorem 6.1. ([Se3], [Se4] and [Se7]). *Let S be as above. If (27) is true with a small enough ϵ, then there is an $\epsilon_1 \leq C\epsilon$ such that*
(28) *the unit normal $n(y)$ to S, pointing towards Ω^+, is in $BMO(S, d\mu)$, with norm $\leq \epsilon_1$.*

Conversely, if (28) is true for a small enough ϵ_1, then there is an $\epsilon \leq C\epsilon_1$ such that (27) is true.

The definition of $BMO(S)$ is not a surprise : we ask for

$$\mu(B(x,r))^{-1} \int_{B(x,r)} | n(y) - n_{x,r} | \, d\mu(y) \leq \epsilon_1, \tag{29}$$

where $n_{x,r} = \mu(B(x,r))^{-1} \int_{B(x,r)} n(y) d\mu(y)$, $x \in S$ and $r > 0$.

Remark 6.2. When $k = 1$, the result says that the Cauchy operator on a rectifiable Jordan curve Γ is almost anti-self adjoint (or the Hardy spaces are almost orthogonal) if and only if Γ is a "chord-arc curve with small constant" (with the definition of II, 8.4, we should say "constant close to 1"). This was essentially proved by Coifman and Meyer [CM2] (also see [Dv0] for the "only if" part).

For this reason, S. Semmes decided to call the surfaces that satisfy (27) (or (28)) "chord-arc surfaces with small constant" (we'll write CASSC's).

Remark 6.3. There are many other equivalent geometric definitions of CASSC's. For instance, (27) implies

(30) for all $x \in S$ and $r > 0$, there is a hyperplane H such that each $y \in S \cap B(x,r)$ is at distance $\leq \epsilon_2 r$ from H ;

(31) for all $x \in S$ and $r > 0$, $| \mu(S \cap B(x,r)) - a_k r^k | \leq \epsilon_3 r^k$, where $a_k r^k$ is the volume of a ball of radius r in $\mathbb{R}^k$;

(32) if $D(x,y)$ denotes the geodesic distance between x and $y \in S$, then $| x-y | \leq D(x,y) \leq (1+\epsilon_4) | x-y |$;

(33) for all $x \in S$ and $r > 0$, there is a surface $\tilde{S}$, which is the image of an ϵ_5-Lipschitz graph by an isometry of $\mathbb{R}^{k+1}$, and such that $\mu(S \cap B(x,r) \cap \tilde{S}^c) \leq \epsilon_5 r^k$.

All these properties are various ways of saying that S looks very much like a hyperplane at all scales. Note that (33) makes it easy to show that C_S is bounded (once (33) is deduced from (28)!). Of course, they are all true for chord-arc curves. One thing we are missing here, however is a very good parametrization (we'll talk about this in Section 10).

Finally, (31) and (32) (with $\epsilon_3 + \epsilon_4$ small enough) imply (27), too.

B - ω-regular surfaces

Let us mention now a generalization of Theorem 4.2 above. The main difficulty of studying surfaces is avoided by assuming we have a very nice parametrization of S.

Definition 6.4. Let $\omega \in A_\infty(\mathbb{R}^k)$ be a weight of the Muckenhoupt class. The function $z : \mathbb{R}^k \to \mathbb{R}^n$ will be called "ω-regular" if there is a constant $C \geq 0$ such that

(34) $$| z'(x) | \leq C\omega(x)^{1/k}$$

and

(35) $$| \{x \in \mathbb{R}^k \ : \ z(x) \in B(w,r)\} |_{\omega dx} \leq Cr^k$$

for all $w \in \mathbb{R}^n$ and $r > 0$.

Comments. The condition (34) is meant "distributionwise". Since ω is a weight in A_∞, it is equivalent to

(36) $$| z(x) - z(y) | \leq C \left(\int_B \omega(u) du \right)^{1/k},$$

where $B = B(x, | x-y |)$, for instance.

The reader will find all needed knowledge on A_∞-weights in [Jé] or [GR]. He can also content himself with the case when $\omega = 1$ (in this case, z is Lipschitz).

If we define a measure μ on $\mathbb{R}^n$ by $\mu(f) = \int_{\mathbb{R}^k} f(z(x))\omega(x)dx$, condition (35) means that $\mu \in \Delta$. Furthermore, (34) (or (36)) implies that $\mu \in \sum$.

Theorem 6.5 [Dv5]. *Let $z \ : \ \mathbb{R}^k \to \mathbb{R}^n$ be ω-regular, and define μ by $\int_{\mathbb{R}^n} f\, d\mu = \int_{\mathbb{R}^k} f \circ z \ \omega\, dx$. Then the operators T^*_μ defined in Section 1 are bounded on $L^p(d\mu)$ for $1 < p < +\infty$.*

The initial proof used Proposition 3.2, but not Corollary 3.6 directly, because the author was not able to find directly "big pieces of Lipschitz graphs" in $S = z(\mathbb{R}^k)$. Instead, one showed that S contains big pieces of something that contains big pieces of .... something that contains big pieces of Lipschitz graphs, and applied a few times an analogue of Corollary 3.6. We'll see a more direct proof in Sections 7 and 8.

Remark. We did not introduce weights in Definition 6.4 merely by thirst of generalizations. First, a converse which is stated in Section 9 (Theorem 9.5) apparently needs weights ω other than 1. Similarly, for some sets S, we know how to find an ω-regular parametrization of S, but we do not know if they admit 1-regular parametrizations (see Sections 9 and 10). Also, the ω-regular mappings have a few interesting properties of their own, in particular in relation with quasiconformal mappings.

One easily checks that if z is ω-regular, then its differential Dz exists almost everywhere, and is quasisymmetric : for each x such that $Dz(x)$ exists, there is a $\lambda > 0$ such that $\lambda \mid v \mid \leq \mid Dz(x) \cdot v \mid \leq C\lambda \mid v \mid$ for all $v \in \mathbb{R}^k$ (C does not depend on x).

Also, if z is ω-regular and h is a quasiconformal mapping of $\mathbb{R}^k$, then $z \circ h$ is $\tilde{\omega}$-regular, where $\tilde{\omega}(x) = \omega(h(x))$.Jacobian$(h)(x)$. ($\tilde{\omega}$ is an A_∞-weight because the Jacobian of h is in A_∞, by a result of F. Gehring [Ge].)

We do not know at this time the precise class of weights ω such that there is an ω-regular mapping z. It is proved in [DS2] that ω must belong to the so-called "strong-A_∞ class", and Semmes recently proved that some condition between A_1 and strong-A_∞ is sufficient [Se11]. See Section 10 for other questions related to this one.

C - Stephen Semmes surfaces

Definition 6.6. Let us denote by $S(k)$ the class of hypersurfaces $S \subset \mathbb{R}^{k+1}$ such that there is a $\mu \in \sum$ with $\operatorname{supp}\mu = S$, and a constant $C \geq 0$ such that

(37) for all $x \in S$ and $r > 0$, there exist points x_1 and x_2, that lie in different connected components of $\mathbb{R}^{k+1} \setminus S$, and such that $\mid x_i - x \mid \leq r$, but $\operatorname{dist}(x_i, S) \geq C^{-1} r$ for $i = 1, 2$.

Examples.

• In dimension $k = 1$, Lipschitz graphs, chord-arc curves are in $S(1)$. The example of Figure 1 is in $S(1)$, but not the example of Figure 2.

• When $k \geq 2$, it is no longer true that all $S \in S(k)$ are of the form $z(\mathbb{R}^k)$ for an ω-regular z. For instance, the topology of S can be quite complicated. A simple example with a few handles is suggested in Figure 3.

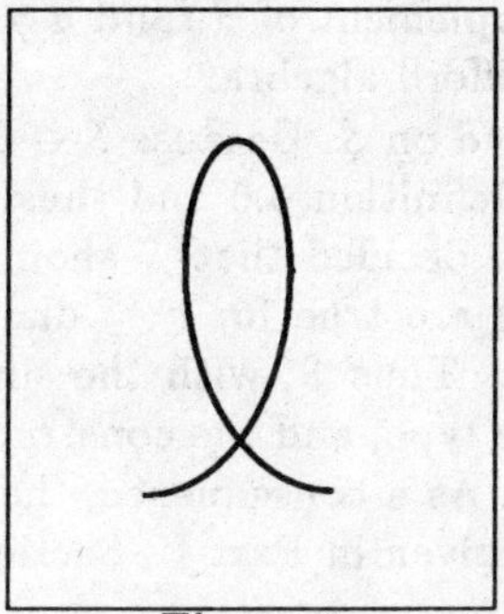

Figure 1

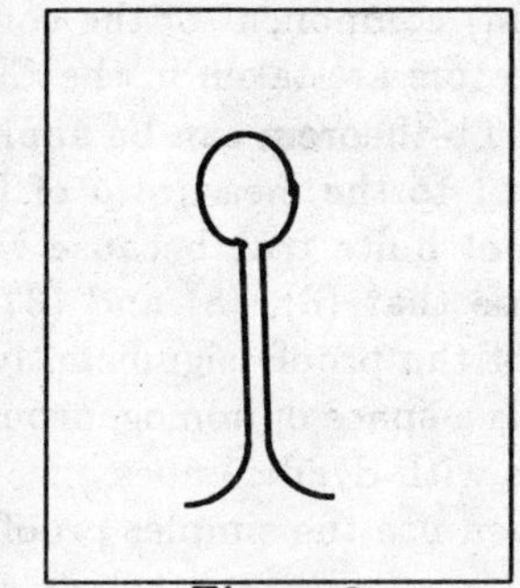

Figure 2

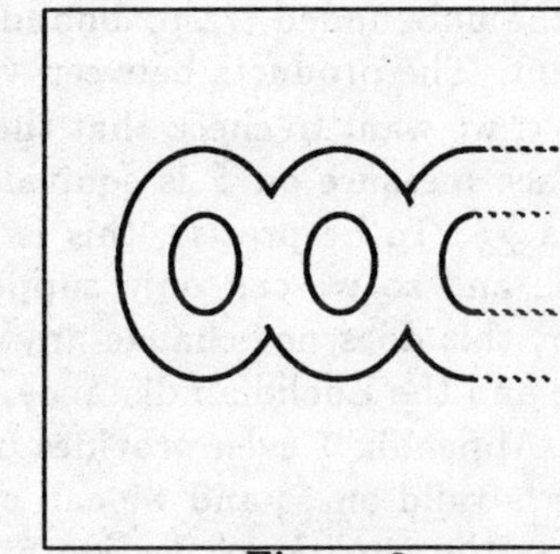

Figure 3

• Also, note that Definition 6.6 does not put any restriction on the number of components of S or its complement.

• A compactness argument shows that the image of $\mathbb{R}^k$ by a bilipschitz mapping from $\mathbb{R}^k$ to $\mathbb{R}^n$ is in $S(k)$ ([Väl] p. 111).

Theorem 6.7 [Se5]. *If $S \in S(k)$ and K is a good kernel, the operator T_μ^* of Section 1 is bounded on $L^p(S, d\mu)$ for $1 < p < +\infty$.*

Actually, Semmes' proof gives a slightly less general result. To describe that proof, we shall make the a priori assumption that S be a smooth, compact submanifold of $\mathbb{R}^{k+1}$, and also the topological assumptions that S be connected, orientable, and that its complement in $\mathbb{R}^{k+1}$ have exactly two connected components. It is easy to get rid of the smoothness assumptions by a limiting argument, but the topological assumptions cannot be dealt with so easily. Also, Semmes'proof does not work for all kernels, but only for a (large enough) class of real-analytic kernels. Theorem 6.7 is still true as we stated it, though, because it can be shown that S CBPLG as soon as $S \in S(k)$ (see Section 7).

Let us now describe Semmes'proof. The idea is quite amusing : apply the Tb-theorem just like when $k = 1$. To do so, one first has to find a substitute for the Cauchy formula. Let us view $\mathbb{R}^{k+1}$ as a subset of $C_k(\mathbb{R})$, the Clifford algebra with k generators over $\mathbb{R}$ (for a precise definition of $C_k(\mathbb{R})$ and its basic properties used below, see [BDS]). Call e_0 the unit, and $e_1, \cdots e_k$ the generators (recall that $e_i^2 = -e_0$ and $e_i e_j = -e_j e_i$ for $i, j \neq 0$ and $i \neq j$). Every point of $\mathbb{R}^k$ is identified with the element $x_0 e_0 + \sum_{i=1}^k x_i e_i$ of $C_k(\mathbb{R})$, and the Clifford-valued analogue of the Cauchy kernel is defined by

$$K(x) = \| x \|^{-k-1} \left\{ x_0 e_0 - \sum_{i=1}^{k} x_i e_i \right\}.$$

Our assumptions on S allow us to integrate by parts, and in particular to use the following consequence of the "Cauchy-Clifford formula" :

$$(38) \qquad \int_S K(x-y) n(y) dy = \int_S n(y) K(x-y) dy = \begin{cases} 0 & \text{if } x \in \Omega^- \\ a & \text{if } x \in \Omega^+, \end{cases}$$

where $n(y)$ is the unit normal to S at y, dy is the surface measure on S, Ω^- (respectively Ω^+) is the unbounded (resp. bounded) component of the complement of S, and $a \neq 0$ is a constant. The products between vectors are taken in the Clifford algebra.

Next, we want to check that the Tb-theorem can be applied on S. Because $S \in S(k)$, the surface measure on S is equivalent to the measure μ of Definition 6.6 and thus is in the class $\sum$. To be precise, this is not quite true because we decided that S should be compact, and so we can only suppose that (3), (8) and (37) are true for $r \leq$ diam S. However, this does not change any of the proofs significantly. Thus S, with the surface measure and the Euclidian distance, is a space of homogeneous type, and the construction given in Appendix 1 even provides us with dyadic cubes on S. As a consequence, the Tb-theorem is valid on S, and we can even use the simpler proof given in Part II, Sections 2 and 3 (see the remarks of Section 5).

The next difficulty is that we want to apply the Tb-theorem with a function b (the unit normal to S) which is no longer complex, but takes Cliford-vector values. Since K is also vector-valued, let us precise some of the notations.

Denote by C_S the principal value operator defined by the antisymmetric kernel $K(x-y)$. The definition of C_S is not a problem : we can use a formula like (4) or the smoothness of S. The operator C_S maps real-valued functions on vector-valued functions, but we can also let it act (on the left or on the right) on vector-valued functions (such as the unit normal to S) : we just take the products of vectors in the Clifford algebra. The result is then a function with values in $C_k(\mathbb{R})$. With these conventions, it follows from (38), and a small limiting argument made easier by our smoothness assumption, that $C_S n = 0$ in BMO, whether C_S acts on the left or on the right.

Let us now check that the unit normal is paraaccretive. That is, let us find constants $C > 0$ and $\delta > 0$ such that, for each $x \in S$ and $0 < r \leq$ diam S, there is a "dyadic cube" $Q \subset S$ such that $\mathrm{dist}(x, Q) \leq Cr$, $\frac{1}{C}r \leq \mathrm{diam}\ Q \leq Cr$, and such that $\left\| \frac{1}{|Q|} \int_Q n(y)dy \right\| \geq \delta$. Of course, we do not want the constants C and δ to depend on the constants in the smoothness assumption, or on the diameter of S. This is similar to the definition of paraaccretivity given in page 35, but we now use a different set of cubes, and modules of complex numbers have been replaced by lengths of vectors.

To prove that n is paraaccretive, pick a point $x \in S$ and a radius $r > 0$, and apply (38) to the points x_1 and x_2 of Definition 6.6. Taking the difference yields

$$\int_S [K(x_1 - y) - K(x_2 - y)]\, n(y)dy = \pm a.$$

Note that the kernel $p(y) = K(x_1 - y) - K(x_2 - y)$ satisfies the same estimates as a Poisson kernel : $\| p(y) \| \leq Cr(r + |x-y|)^{-k-1}$ and $\| \nabla p(y) \| \leq Cr^2(r + |x-y|)^{-k-2}$. Fix a large constant $\lambda > 0$, and cover S by disjoint cubes Q_i of diameters comparable to $\lambda^{-1}r$, and then replace $p(y)$ by $\tilde{p}(y)$, where $\tilde{p}(y)$ is obtained as follows. Let $Q_i(y)$ be the cube Q_i that contains y, and let $\tilde{p}(y)$ be the mean value of p on $Q_i(y)$ if $\mathrm{dist}(x, Q_i(y)) \leq \lambda r$, and 0 otherwise. If λ is large enough, $\tilde{p}$ is so close to p that one still has $\left\| \int_S \tilde{p}(y) n(y) dy \right\| \geq \left| \frac{a}{2} \right|$. The paraaccretivity of n then follows by replacing $\tilde{p}$ by one of its coordinates, and then choosing one of the Q_i's on which the integral is largest.

Our last task is to modify the proof of Part II, Sections 2 and 3, to make it work in a Clifford-valued context. We still want to use the same computations, but we shall have to be careful because Clifford vectors do not commute. The delicate part of the proof is the construction of the Riesz basis : the estimates for $C_{QQ'}$ will essentially be the same.

As in the general paraaccretive case, we first modify our set of cubes in such a way that $\left\|\frac{1}{|Q|}\int_Q b(y)dy\right\| \geq \delta$ for each dyadic cube Q. Note that a non-zero vector of $\mathbb{R}^{k+1}$ is always invertible in $C_k(\mathbb{R})$: its inverse is its "conjugate", divided by the square of its norm. Thus, $\int_Q b(y)dy$ is invertible, and its inverse is a Clifford vector with length $\leq C\delta^{-1} \mid Q \mid^{-1}$. We shall replace the projection operators of Part II by two operators F_k and F_k^* defined by $F_k f(x) = \left[\int_Q f(t)b(t)dt\right]\left[\int_Q b(t)dt\right]^{-1}$ and $F_k^* f(x) = \left[\int_Q b(t)dt\right]^{-1}\left[\int_Q b(t)f(t)dt\right]$, where Q is, as before, the dyadic cube of size 2^{-k} that contains x.

If we set $\Delta_k = F_{k+1} - F_k$ and $\Delta_k^* = F_{k+1}^* - F_k^*$, the same computations as before yield (13) and (14) for F_k and Δ_k, and F_k^* and Δ_k^*. The identity (15) has to be modified a little : one writes $\int \Delta_k u(x)b(x)\Delta_\ell^* v(x)dx = 0$ instead. Lemma 2.1 also stays true for both Δ_k and Δ_k^* : the first inequality is proved as before and, for the second inequality, one modifies the duality argument and writes

$$\| f \|_2^2 = \int f bb^{-1} f = \sum_k \sum_\ell \int\int (\Delta_k f) b(\Delta_\ell(b^{-1}f)) = \sum_k \int\int (\Delta_k f) b(\Delta_\ell(b^{-1}f)).$$

The rest of the estimate is as before. Note that the argument also works when f has values in $\mathbb{R}^{k+1}$.

The conclusion of these computations is that every real-valued function can be decomposed as $f = \sum_k \Delta_k f$, with $\| f \|^2 \sim \sum \| \Delta_k f \|^2$. We can decompose f even further wy writing $\Delta_k f = \sum_Q \alpha_Q f_Q$, where the sum is over all dyadic cubes of size 2^{-k}, $\alpha_Q = \| \mathbb{1}_Q \Delta_k f \|_2$, and $f_Q = \alpha_Q^{-1}\mathbb{1}_Q \Delta_k f$. This yields the decomposition $f = \sum \alpha_Q f_Q$, where this time Q runs over all dyadic cubes, the coefficients α_Q are such that $\| f \|^2 \sim \sum \mid \alpha_Q \mid^2$, and the functions f_Q have the following properties. First, f_Q is supported on Q, and is constant on each of the cubes of the next generation. As a consequence, $\| f_Q \|_\infty \leq C \mid Q \mid^{-1/2} \| f_Q \|_2 \leq C \mid Q \mid^{-1/2}$. Finally, f_Q satisfies the cancellation condition $\int f_Q(x)b(x)dx = 0$ because $\int_Q \Delta_k f(x)b(x)dx = 0$ for every cube Q of size 2^{-k}. Note the we could have used the Δ_k^* and obtained a similar decomposition, but with the cancellation condition $\int b(x) f_Q(x)dx = 0$.

At this stage, it would probably be possible to go on as in Part II, and express the functions f_Q in terms of some Riesz basis. The author of these notes is far too afraid of things that do not commute to proceed in this direction, and so we shall keep our functions f_Q. This is not too bad, because they satisfy exactly the same estimates as the h_Q^ϵ's of Part II, Section 2.

Let us come back to our operator C_S, and let us show that the operator $T = M_b C_S M_b$ is bounded on L^2. The boundedness of C_S will of course follow, since b is invertible. Because we want to consider test-functions which are vector-valued, we shall use the notation $< fT, g >$ to denote the effect on the effect on g (by an action on the left) of the image of f by T (acting on the right). In other words, if K were integrable, we would have $< fT, g > = \int\int f(y)b(y)K(x-y)b(x)g(x)dx\, dy$.

We now use the Δ_k's to write $f = \sum \alpha_Q f_Q$, and the Δ_k^*'s to write $g = \sum \beta_Q g_Q$, so that

$$< fT, g > = \sum_Q \sum_{Q'} \alpha_Q \beta_{Q'} < f_Q T, g_{Q'} > .$$

The coefficients $C_{QQ'} = < f_Q T, g_{Q'} >$ can be estimated like in Part II, Section 3, using the standard estimates on K, the properties of dyadic cubes, and the facts that $T1 = T^t 1 = 0$ and $\int f_Q b = \int b g_Q = 0$. The result is that the matrix with coefficients $\| C_{QQ'} \|$ defines a bounded operator on ℓ^2, and that

$$\| < fT, g > \| \leq C \left(\sum | \alpha_Q |^2 \right)^{1/2} \left(\sum | \beta_Q |^2 \right)^{1/2} \leq C \| f \| \| g \| .$$

This proves the boundedness of C_S.

As we said earlier, the proof of S. Semmes also works for other kernels than the Cauchy-Clifford kernel K. The idea is to use the Tb-theorem again, and to compute the image of the unit normal by an integration by parts. If the result is the image of a bounded function by an operator we already know how to treat, we are in business. The details can be found in [Se5].

Exercise. Complete the proof of the Tb-theorem in the case of a function b with values in $\mathbb{R}^{k+1} \subset C_k(\mathbb{R})$. Use a paraproduct with a kernel of the form

$$P(x, y) = \sum_Q \alpha_Q f_Q(x) \theta_Q(y)$$

(where the f_Q's are obtained by decomposing a function of BMO with the Δ_k^*'s), or of the form

$$P(x, y) = \sum_Q \alpha_Q \theta_Q(x) f_Q(y)$$

(where the f_Q's are obtained with the Δ_k's).

7. Finding big pieces : clouds and shadows

Let us come back to the good-λ method. We are given a "surface" $S \subset \mathbb{R}^n$, and a measure $\mu \in \sum$ supported on S, and we wish to apply Corollary 3.6, so we wish to show that S CBPLG.

A first step in that direction is to show that S "has big projections" in the following sense : there is a $\theta > 0$ such that for each $x \in S$ and $r > 0$, there is a k-dimensional vector space V such that, if π is the orthogonal projection on V, then

$$(39) \qquad | \pi(S \cap B(x, r)) | \geq \theta r^k$$

(where | | denotes the Lebesgue measure on V).

It is clear that, if $\mathcal{S}$ CBPLG, then $\mathcal{S}$ "has big projections" : if $E \subset \mathcal{S} \cap B(x,r)$ is contained in the graph of an M-Lipschitz function from V to $V^\perp$, then $|\pi(E)| \geq C^{-1}(M+1)^{-1}\mu(E)$.

In most known cases, showing that $\mathcal{S}$ has big projections is the easy part. For instance, if $\mathcal{S}$ is in $S(k)$ (see Definition 6.6), we can take for V the hyperplane orthogonal to the line L that contains the points x_1 and x_2 of Definition 6.6. Indeed, if L' is any line that meets $B(x_1, C^{-1}r)$ and $B(x_2, C^{-1}r)$, then $L' \cap \mathcal{S} \cap B(x, 2r)$ is not empty. From this, we deduce that $\pi(\mathcal{S} \cap B(x,2r))$ contains a ball of radius $C^{-1}r$ centered at $\pi(x_1)$, which gives the "big projection" for $2r$ (see the picture).

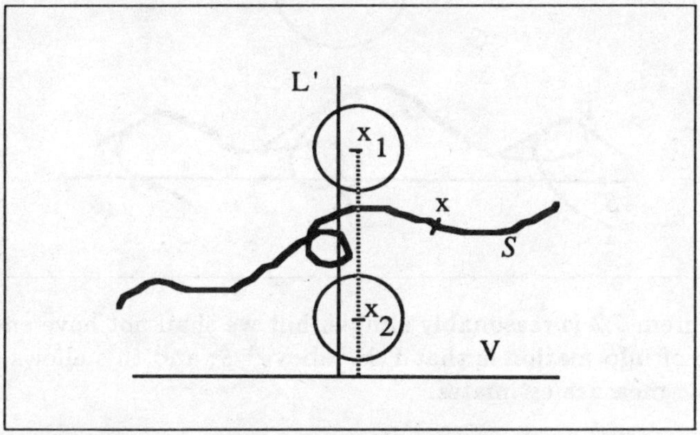

Let us assume now that we proved that $\mathcal{S}$ "has big projections". We still have to find a piece of Lipschitz graph in $\mathcal{S} \cap B(x,r)$ or, equivalently, a subset $E \subset \mathcal{S} \cap B(x,r)$ such that $\pi_{|E}$ is bilipschitz (the "Clouds and shadows problem").

A condition is given in [Dv6], that allows in certain cases to solve the "Clouds and shadows problem". The condition is not very pleasant, and the proof by a stopping time argument is not too simple, so we'll just mention a corollary :

Theorem 7.1. *If $\mathcal{S} = z(\mathbb{R}^k)$ for some ω-regular function $z : \mathbb{R}^k \to \mathbb{R}^n$, or if $\mathcal{S} \in S(k)$, then $\mathcal{S}$ CBPLG (see Definitions 6.4, 6.6 and 3.4 for the jargon).*

This theorem gives a new proof of Theorems 6.5 and 6.7. One can also define higher codimension analogues of $S(k)$ for which Theorem 7.1 also holds (see [Dv6]). We shall see in Section 8 a faster proof when $\mathcal{S} = z(\mathbb{R}^k)$ for an ω-regular z. In the case of $S(k)$, there is another proof, too ([DJe]), which also gives a little more :

Theorem 7.2. *Let $\mathcal{S} \in S(k)$. Assume that $O \in \mathcal{S}$, and that the ball B with center $(0,\cdots,0,1)$ and radius C^{-1} is entirely contained in one of the connected components of $\mathbb{R}^{k+1}\backslash\mathcal{S}$ (call this component $\mathcal{O}$). Then there is a constant $C_1 \geq 0$, that depends only on C and the constants of Definition 6.6, and a C_1-Lipschitz function A, defined on $B(0,(2C)^{-1}) \subset \mathbb{R}^k$, and such that the graph Γ of A satisfies :*

$$\Gamma \subset \bar{\mathcal{O}} \tag{40}$$

and

(41) $$\mu(\Gamma \cap S) \geq C_1^{-1}$$

(see the picture).

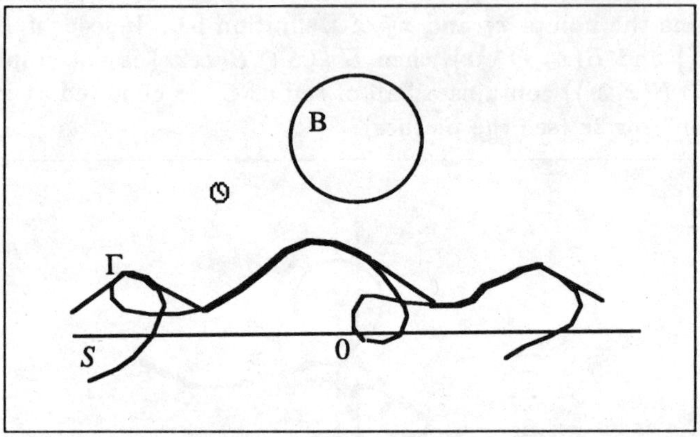

The proof of Theorem 7.2 is reasonably simple, but we shall not have enough time to give it. The new piece of information is that Γ is "above" S, and this allows one to obtain the following harmonic measure estimates.

Corollary 7.3. *Let $S \in S(k)$, and O be one of the components of $\mathbb{R}^{k+1} \backslash S$. If O is a Non-Tangential Access domain, then the harmonic measure on ∂O (relative to O) is in the Muckenhoupt class A_∞ with respect to surface measure on S.*

We refer to [DJe] or [JK] for the definition of NTA domains.

Let us mention that S. Semmes has another proof of Corollary 7.3, obtained independently by a method related to the "corona construction" [Se9]. This method had been used by P. Jones and S. Semmes to treat the case of the image of a bilipschitz mapping from $\mathbb{R}^k$ into $\mathbb{R}^{k+1}$. Semmes obtained the special case of Corollary 7.3 when both sides of S are NTA at the same time as [DJe], and the general case a little later.

The condition $S \in S(k)$ is not necessary for the proofs of Theorem 7.2 and Corollary 7.3 ; in particular, the conclusion of the corollary is still true if the k-dimensional Hausdorff measure on S is in $\sum$, if for each $x \in S$ and $r > 0$, there are two k-dimensional disks D_1 and D_2, with radius $\frac{r}{C}$, at distance $\leq r$ from x, and that are contained in different components of $\mathbb{R}^{k+1} \backslash S$ (of course, the NTA condition on Harnack chains is still needed, too).

The idea of the proof is to compare harmonic measure on O with harmonic measure on a Lipschitz domain $\Omega \subset O$ such that $| \partial\Omega \cap \partial O |$ is relatively large. One uses the theorem to find Ω, and techniques of [JK] to conclude. More details can be found in [DJe].

8. Bilipschitz mappings inside Lipschitz functions

The following theorem of P. Jones will allow us to solve the "clouds and shadovs problem" for regular mappings.

Theorem 8.1 [Jn2]. *Let I be the unit cube of $\mathbb{R}^k$ and $f : I \to \mathbb{R}^k$ be a Lipschitz function. For each $\epsilon > 0$, there is an integer $M = M(\| \nabla f \|_\infty, \epsilon)$, a constant $C(\| \nabla f \|_\infty, \epsilon)$ and sets $E_1, \cdots E_M \subset I$ such that*

$$| f(x) - f(y) | \geq C(\| \nabla f \|_\infty, \epsilon)^{-1} \, | x - y | \tag{42}$$

whenever x and y are in the same E_j, and

$$| f(I \cap E_1^c \cap \cdots \cap E_M^c) | \leq \epsilon. \tag{43}$$

We shall give a proof of this theorem, but let us see how it implies that the image of a 1-regular mapping always contains big pieces of Lipschitz graphs.

Corollary 8.2. *If $f : I \to \mathbb{R}^k$ is Lipschitz, and if $| f(I) | \geq \delta$, then there is a compact set $E \subset I$ such that $| E | \geq \theta(\| \nabla f \|_\infty, \delta) > 0$ and*

$$| f(x) - f(y) | \geq C(\| \nabla f \|_\infty, \delta)^{-1} \, | x - y | \quad \text{for } x, y \in E.$$

The corollary clearly follows from the Theorem (choose $\epsilon = \frac{\delta}{2}$). A first proof of the Corollary was given in [Dv6], but it is a little less engaging.

Let us now use the corollary to prove that if $z : \mathbb{R}^k \to \mathbb{R}^n$ is 1-regular (i.e., ω-regular with the weight $\omega = 1$), then $S = z(\mathbb{R}^k)$ CBPLG. The general result (when ω is any A_∞-weight) would require slightly different versions of Theorem 8.1 and Corollary 8.2, which are both true (and can be deduced easily from these results), but we decided to simplify the statements and take $\omega = 1$. Let $w \in S$ and $r > 0$ be given.

Choose a point $x_0 \in \mathbb{R}^k$ such that $z(x_0) = w$. If C is large enough, the cube Q centered at x_0 and with sidelength $\frac{r}{C}$ is such that $z(Q) \subset B(w, r)$.

Lemma 8.3. *There is a k-dimensional subspace V of $\mathbb{R}^n$ such that, if π is the orthogonal projection onto V, $| \pi(z(Q)) | \geq \delta \, | Q |$.*

Of course, we do not want δ to depend on Q. If δ were allowed to depend on Q, the lemma would be completely trivial : one can find a point x in the interior of Q, such that z is differentiable at x ; then the derivative of z at x is of rank k (because z is 1-regular), and if V is chosen well, $\pi(z(Q))$ will contain a small ball around $\pi(z(x))$ (by elementary degree theory). The fact that we can choose δ independent of Q and z (provided the regularity constants of z stay bounded) now follows from a simple compactness argument, which is left as an exercise (take a sequence (z_n, Q_n) for which the best δ_n tends to 0, extract a convergent subsequence after normalization, and use the argument above and some more elementary degree theory). More details can be found in [Dv6], lemme 10, p. 96.

Apply Corollary 8.2 to the function $f(x) = \pi(z(x))$ (possibly after an affine change of variables if Q is not the unit cube). We get a set $E \subset Q$, with $| E | \geq \theta | Q |$, and such that $\pi \circ z$ is bilipschitz on E. The set $\tilde{E} = z(E)$ satisfies $\mu(\tilde{E}) \geq \theta r^k / C$, and $\pi_{|\tilde{E}}$ is also bilipschitz, so $\tilde{E}$ is the piece of Lipschitz graph we were looking for.

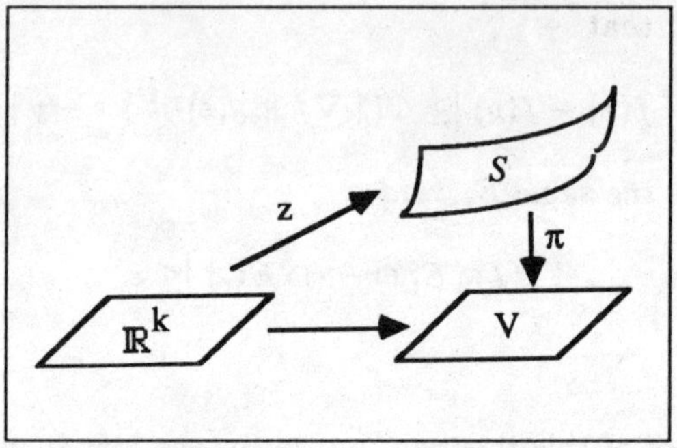

Let us now prove Theorem 8.1. We shall need a way to control the situations where $| f(x) - f(y) | \leq \delta | x - y |$ ($\delta > 0$ is a small constant, to be chosen later). The general idea is that, in such a case, either $| f(B) |$ is very small for some ball B containing x and y, or else f is not close to being affine on B, and so f'' must be large somewhere in B. We intend to prove that the second possibility cannot occur very often by associating to B a large "wavelet coefficient" for f'.

To simplify notations, let us assume that f is 1-Lipschitz, and also is defined on the whole $\mathbb{R}^k$. We need a few notations before we start the stopping time argument.

Let us fix a constant $C \geq 1$; we'll say that two dyadic cubes Q_1 and $Q_2 \subset I$ are "semi-adjacent" if $| Q_1 | = | Q_2 |$, and if

$$| Q_1 |^{1/k} \leq \operatorname{dist}(Q_1, Q_2) \leq C | Q_1 |^{1/k} .$$

Also, if Q is a dyadic cube, the father of Q (i.e., the only dyadic cube of sidelength $2 | Q |^{1/k}$ which contains Q) will be denoted by Q^*.

Finally, let us choose a wavelet basis (like in Part I), and let us say for convenience that we use compactly supported wavelets of class C^4 (many other things would work just as well). Let us write $\nabla f = \sum_Q \sum_\epsilon a_Q^\epsilon \psi_Q^\epsilon$ the wavelet expansion of ∇f in the basis (ψ_Q^ϵ) (note that the a_Q^ϵ's are vector-valued).

The following lemma will be used to control the number of places where f is clearly far from being affine.

Lemma 8.4. *If the constant $C_0 \geq 0$ is large enough, then the following result is true. Let Q_1 and Q_2 be semi-adjacent dyadic cubes, and suppose that, for some $\alpha > 0$, $| f(Q_1) | \geq \alpha | Q_1 |$. Also suppose that $| f(x_1) - f(x_2) | \leq C_0^{-1} \alpha | x_1 - x_2 |$ for some $x_1 \epsilon Q_1$ and $x_2 \in Q_2$. Then, there is a dyadic cube $Q \subset C(\alpha) Q_1$, with $| Q | \geq C(\alpha)^{-1} | Q_1 |$, and such that $| a_Q^\epsilon | \geq C(\alpha)^{-1} | Q_1 |^{1/2}$ for some ϵ.*

This lemma can be proved directly, but it is just as simple for us to use a compactness argument. If Lemma 8.4 is false, there is a sequence f_n of 1-Lipschitz functions, such that the hypotheses are satisfied for the unit cube Q_1 and some semi-adjacent Q_2 (and the same $\alpha > 0$), but such that $| a_Q^\epsilon | \leq \frac{1}{n}$ for all Q of size $\geq 2^{-n}$ contained in $2^n Q_1$. Extract a subsequence $f_{n(i)}$ that converges uniformly to a function f_∞, and such that the $\nabla f_{n(i)}$ also converge (locally) in all L^p, $p < +\infty$. The wavelets coefficients of ∇f_∞ are all 0, and so f_∞ is affine. Since f_∞ is 1-Lipschitz, and there are points $x_1 \in Q_1$ and $x_2 \in Q_2$ such that $| f_\infty(x_1) - f_\infty(x_2) | \leq 2C_0^{-1}\alpha | x_1 - x_2 |$, one gets $| f_\infty(Q_1) | \leq CC_0^{-1}\alpha < \frac{\alpha}{2} | Q_1 |$ (if C_0 is chosen correctly).

On the other hand, we know that $| f_n(Q_1) | \geq \alpha | Q_1 |$ for all n, and this implies that $| f_\infty(Q_1) | \geq \alpha | Q_1 |$; the ensuing contradiction proves the lemma.

Exercise. Prove what was just said. Hint : if $| f_\infty(Q_1) |$ is small, then $f_\infty(Q_1)$ is contained in an open set with small measure, and the same must be true of $f_n(Q_1)$ for n large enough.

Remark. We decided to use wavelets, but there are many other options : about any result giving some control on how the function f is approximated by affine functions at all scales would probably do as well.

Next, let us introduce two categories of cubes where something unpleasant happens.

Call $\mathcal{E}_1$ the set of dyadic cubes $Q_1 \in I$ such that there is a semi-adjacent cube Q_2, for which $| f(Q_1) | \geq \epsilon/2 | Q_1 |$, but such that there are points $x_1 \in Q_1$ and $x_2 \in Q_2$ with $| f(x_1) - f(x_2) | \leq C_0^{-1}\epsilon | x_1 - x_2 | /2$ (the number ϵ is the same as in the theorem).

Call $\mathcal{E}_2$ the set of dyadic cubes $Q \subset I$ such that $| f(Q) | \leq \frac{\epsilon}{2} | Q |$.

We can throw away all the cubes of $\mathcal{E}_2$, because $| f(\bigcup_{\mathcal{E}_2} Q) | \leq \frac{\epsilon}{2}$. We also intend to throw away the points of I that belong to too many cubes of $\mathcal{E}_1$.

We know from Lemma 8.4 that if $Q_1 \in \mathcal{E}_1$, there is a wavelet coefficient a_Q^ϵ, where Q is not too far from Q_1 for the hyperbolic distance, which is $\geq | Q_1 |^{1/2} /C$. Therefore, $\sum_{Q_1 \in \mathcal{E}_1} | Q_1 | \leq C \sum_{Q \subset CI} | a_Q^\epsilon |^2 \leq C$ (since $\| \nabla f \|_\infty \leq 1$). If we choose N large enough, we'll get $| B | \leq \frac{\epsilon}{2}$, where $B = \{x \in I \ : x$ belongs to more than N cubes of $\mathcal{E}_1\}$. Since f is 1-Lipschitz, $| f(B) | \leq \frac{\epsilon}{2}$.

Call G the complement in I of $B \cup \left(\sum_{Q \in \mathcal{E}_2} Q\right)$. We just have to split G into M sets $E_1, \cdots E_M$ on which f will be bilipschitz.

We want to associate, to each cube $Q \not\subset \sum_{Q' \in \mathcal{E}_2} Q'$ a certain sequence $\alpha(Q)$ of 0's and 1's ; the length of the sequence will be called $\ell(Q)$.

We want to define $\alpha(Q)$ by induction, starting from I, and then defining $\alpha(Q)$, for each Q, in terms of $\alpha(Q^*)$ (Q^* is the father of Q). First, let us introduce more notations. For each $n \geq 0$, call $\mathcal{A}_n$ the set of all dyadic cubes $Q \in \mathcal{E}_1$ of sidelength 2^{-n} that are not contained in any cube of $\mathcal{E}_2$. Call P_n the set of all (unordered) paris (Q_1, Q_2), where Q_1 and Q_2 are semi-adjacent cubes of $\mathcal{A}_n$; choose any order on P_n, and call $P_{n,1}$, $P_{n,2}$, $\cdots P_{n,\ell} \cdots$

the elements of P_n. We are now ready to define $\alpha(Q)$ for Q of sidelength 2^{-n}, assuming this was done for cubes of sidelength $2^{-(n-1)}$ (we take for $\alpha(I)$ the empty sequence).

Case 1. If $Q \notin \mathcal{E}_1$, and is not contained in any cube of $\mathcal{E}_2$, we simply set $\alpha(Q) = \alpha(Q^*)$.

Otherwise, we will define (or change) $\alpha(Q)$ as follows. We take the set P_n in order, and define (or modify) $\alpha(Q_1)$ and $\alpha(Q_2)$ (where $p_{n,\ell} = (Q_1, Q_2)$) successively for $\ell = 1, \cdots$ Suppose we already went through $p_{n,1}, \cdots p_{n,\ell-1}$, and let $p_{n,\ell} = (Q_1, Q_2)$, where Q_1 and Q_2 are semi-adjacent cubes of $\mathcal{E}_1$. The cubes Q_1^* and Q_2^* are not contained in any cube of $\mathcal{E}_2$, and so $\alpha(Q_1^*)$ and $\alpha(Q_2^*)$ are well defined.

Case 2. First suppose that $\ell(Q_1^*) = \ell(Q_2^*)$.

A - If $\alpha(Q_1^*) \neq \alpha(Q_2^*)$, we leave $\alpha(Q_1) = \alpha(Q_1^*)$ and $\alpha(Q_2) = \alpha(Q_2^*)$.

B - If $\alpha(Q_1^*) = \alpha(Q_2^*) = (\epsilon_1, \cdots \epsilon_{\ell(Q_1^*)})$, we set $\alpha(Q_1) = (\epsilon_1, \cdots \epsilon_{\ell(Q_1^*)}, 0)$ and $\alpha(Q_2) = (\epsilon_1, \cdots \epsilon_{\ell(Q_1^*)}, 1)$.

Case 3. Now suppose that $\ell(Q_1^*) \neq \ell(Q_2^*)$. If $\ell(Q_1^*) > \ell(Q_2^*)$, $\alpha(Q_1^*) = (\epsilon_1, \cdots \epsilon_{\ell(Q_1^*)})$ and $\alpha(Q_2^*) = (\epsilon'_1, \cdots \epsilon'_{\ell(Q_2^*)})$, we keep $\alpha(Q_1) = \alpha(Q_1^*)$, and set $\alpha(Q_2) = (\epsilon'_1, \cdots \epsilon'_{\ell(Q_2^*)}, \epsilon')$,where $\epsilon' \neq \epsilon_{\ell(Q_2^*)+1}$. If $\ell(Q_1^*) < \ell(Q_2^*)$, we exchange Q_1 and Q_2, and do as above.

If the same cube Q appears in more than one pair $p_{n,\ell}$, we do not exactly do as above, but replace $\alpha(Q^*)$ by the last defined value of $\alpha(Q)$ (so, most of the time, $\alpha(Q)$ will change a few times during the definition process).

Note that, if $Q \subset Q'$, then the sequence $\alpha(Q)$ starts with the sequence $\alpha(Q')$. Also, if x is in the good set G, and $I \supset Q^{(1)} \cdots \supset Q^{(n)} \supset \cdots$ is the sequence of dyadic cubes containing x, then the sequence $\alpha(Q^{(n)})$ is well defined because $x \notin \bigcup_{\mathcal{E}_2} Q$, and changes less than CN times during the whole definition process (it only changes for those n's such that $Q^{(n)} \in \mathcal{E}_1$, and even for those n's, it changes less than C times because there is $\leq C$ pairs of semi-adjacent cubes containing $Q^{(n)}$). Therefore, for each $x \in G$, the limit of the $\alpha(Q^{(n)})$'s exists, and has length $\leq CN$. Call $\alpha(x)$ this limit.

Let us call E the set of all sequences of length $\leq CN$, composed of 0's and 1's. For each $i \in E$, let $F_i = \{x \in G : \alpha(x) = i\}$, and let us check that, for each i,

$$(44) \qquad |f(x) - f(y)| > C_0^{-1} \frac{\epsilon |x - y|}{2} \quad \text{for } x, y \in F_i.$$

Let $x, y \in F_i$ for some i, and call Q_1, Q_2 two semi-adjacent cubes such that $x \in Q_1$ and $y \in Q_2$ [such cubes exist, provided the constant C in the definition of semi-adjacent cubes was large enough]. Since $\alpha(x) = \alpha(y)$, the sequence $\alpha(Q_1)$ is the beginning of $\alpha(Q_2)$ if it is not longer, and otherwise $\alpha(Q_2)$ is the beginning of $\alpha(Q_1)$. We made sure, when defining $\alpha(Q_1)$ and $\alpha(Q_2)$, that this would never happen if Q_1 and Q_2 are in $\mathcal{E}_1$. So Q_1 (or Q_2) is not in $\mathcal{E}_1$, and this implies (44) by definition of $\mathcal{E}_1$. This concludes the proof of Theorem 8.1; because the number of sets F_i is less than $CN2^{CN}$.

Exercise. State and prove an analogue of Theorem 8.1, where Lipschitz functions are replaced by functions satisfying (34) (or (36)) for some weight $\omega \epsilon A_\infty$. Deduce a proof of the fact that $z(\mathbb{R}^k)$ CBPLG when z is ω-regular.

Remark. The proof above has the advantage of using very little of the structure of $\mathbb{R}^k$. This is used in [DS4] to extend Theorem 8.1 to some functions defined on a regular set, and to give one more proof of the fact that every $S \in S(d)$ contains big pieces of Lipschitz graphs.

9. Square functions, geometric lemma and the corona construction

In this section, we wish to take advantage of a small delay in the preparation of the manuscript to present a few more results (some of them are actually posterior to the lectures). We shall not have time to see any proof, but we'll try to introduce a few notions that are relevant to our study of singular integrals on subsets of $\mathbb{R}^n$.

A - Square function estimates

Let $\mu \in \sum$, and $S \subset \mathbb{R}^n$ be the support of μ. An easy argument using Rademacher functions shows that, if all the good kernels K of Definition 1.1 give operators T_μ^* that are bounded on $L^2(d\mu)$, then the following square function estimates hold. Let Φ be the class of all odd functions $\psi \in C_c^\infty(\mathbb{R}^n)$; then for each $\psi \in \Phi$, there is a constant $C = C_\psi$ such that

$$\text{(45)}\qquad \begin{aligned} &\int_S \sum_{j\in\mathbb{Z}} \left| \int_S 2^{-jk} \psi\left(\frac{x-y}{2^j}\right) f(y) d\mu(y) \right|^2 d\mu(x) \\ &\leq C \int_S |f(y)|^2 d\mu(y) \quad \text{for all } f \in L^2(d\mu). \end{aligned}$$

If, instead of asking for oddness, we had defined the good kernels by the condition (2), we would have the square function estimate (45) with the slightly larger class $\Phi = \{\psi \in C_c^\infty(\mathbb{R}^n) \ : \ \int_{\mathbb{R}} \psi(t\theta) \mid t \mid^{d-1} dt = 0 \text{ for all } \theta \in S^{n-1}\}$ instead. See [DS3] for more details.

We shall see later that (45) implies the boundedness of all T_μ^*'s in return. The interest of (45) is that it is slightly easier to use, because it can be stated in terms of Carleson measures, like many of the properties that will be mentioned in this section. Also, it is closer to a concept dear to S. Semmes : "Littlewood-Paley theory on the set S".

Before we switch to a different notion, let us mention that Semmes proved slightly different square function estimates, for surfaces $S \in S(k)$ (see Definition 6.6).

Theorem 9.1 [Se5]. *Let $S \in S(k)$, $K(x) = x^* \parallel x \parallel^{-k-1}$ be the Cauchy-Clifford kernel, and $Cf(x) = \int_S K(x-y)n(y)f(y)dy$ be the Cauchy-Clifford integral of f (it is defined for*

$f \in L^2(S)$ *and* $x \notin S$). *Then*

$$(46) \qquad \int_{\mathbb{R}^{k+1}} |\nabla(Cf)(x)|^2 \, dist(x,S) ds \leq C \int_S |f|^2 .$$

The proof of (46) uses the same ideas as the proof of the boundedness of the Cauchy-Clifford operator on S. A simpler proof can be found in [Se6], and variants and applications can be found in [Se8].

B - P. Jones' geometric lemma

For $x \in S \subset \mathbb{R}^n$ and $r > 0$, define $\beta(x,r)$ by

$$(47) \qquad \beta(x,r) = \inf_P \sup_{y \in S \cap B(x,r)} r^{-1} \mathrm{dist}(y,P),$$

where the infimum is taken over all k-dimensional affine subspaces P of $\mathbb{R}^n$.

The number $\beta(x,r)$ measures how well S can be approximated, near x and at the scale r, by k-planes. If S is a Lipschitz graph, for instance, a very brutal estimate only gives $\beta(x,r) \leq C < 1$, but the point is that, for most couples (x,r), $\beta(x,r)$ is much smaller than that. The $\beta(x,r)$'s were introduced in [Jn1], to obtain a quite interesting new proof of the boundedness of the Cauchy integral operator on Lipschitz graphs and Ahlfors-regular curves. In particular, he used the following estimate (now known as the "geometric lemma") to compare S to straight lines.

Theorem 9.2 [Do], [Jn1]. *If S is a Lipschitz curve in $\mathbb{R}^2$, then $\beta(x,r)^2 \frac{d\mu(x)dr}{r}$ is a Carleson measure on $S \times \mathbb{R}_+$, i.e.*

$$(48) \qquad \int_{x \in S \cap B(X,R)} \int_{0 < r \leq R} \beta(x,r)^2 \frac{d\mu(x) dr}{r} \leq CR$$

for all $X \in S$ and $R > 0$.

This theorem is actually an easy consequence of a result of Dorronsoro on affine approximations of functions. P. Jones' main contribution is not the fact that he re-discovered it, but that he had the idea to use it in the context of singular integrals on curves. His idea of measuring the regularity of S with the function $\beta(x,r)$ (again, some sort of "geometric Littlewood-Paley theory") has been the source of a lot of recent work (see Theorem 9.4 and 9.5 below, for instance).

In higher dimensions, it was noted by X. Fang [Fn] that the analogue of (48) does not always hold for Lipschitz graphs. Fortunately, only minor modifications are needed. For $1 \leq p < +\infty$, let

$$(49) \qquad \beta_p(x,r) = \inf_P \{ r^{-k} \int_{S \cap B(x,r)} [r^{-1} \mathrm{dist}(y,P)]^p d\mu(y) \}^{1/p} .$$

be the L^p-analogue of $\beta(x,r) = \beta_\infty(x,r)$, where the inf is still taken over all k-planes. The geometric lemma generalizes as follows.

Theorem 9.3 [Do], [Fn]. *If S is a k-dimensional Lipschitz graph in $\mathbb{R}^n$, and if $1 \le p < \frac{2k}{k-2}$, then $\beta_p(x,r)^2 \frac{d\mu(x)dr}{r}$ is a Carleson measure on $S \times \mathbb{R}^+$, which means that*

(50)
$$\int_{x\in S\cap B(X,R)} \int_{0<r\le R} \beta_p(x,r)^2 \frac{d\mu(x)dr}{r} \le CR^k$$

for all $X \in S$ and $R > 0$.

We shall state, in a later subsection, a converse to Theorem 9.3. In the mean time, let us say a few words about P. Jones' traveling salesman theorem. The question is to determine for which subsets S of $\mathbb{R}^2$ it is possible to find a curve with finite length, that passes through all points of S. The answer is as follows.

Theorem 9.4 [Jn3]. *Let S be a compact subset of $\mathbb{R}^2$, with finite 1-dimensional Hausdorff measure. There exists a curve $\Gamma \subset \mathbb{R}^2$ with finite length, and such that $S \subset \Gamma$, if and only if*

(51)
$$\int_{\mathbb{R}^2} \int_0^1 \beta(x,r)^2 \frac{dx\,dr}{r^2} < +\infty,$$

where $\beta(x,r)$ is defined as in (47) even when $x \notin S$, and is set equal to 0 when $S \cap B(x,r)$ is empty.

P. Jones also characterizes the sets $S \subset \mathbb{R}^2$ that are contained in an Ahlfors-regular curve : the condition (51) is then replaced by an appropriately scale-invariant Carleson measure condition. See [Jn3] for more details, and [BJ], say, for a nice application to harmonic measure.

When $S = \operatorname{supp}\mu$ for some $\mu \in \sum$ (or, equivalently, if the restriction to S of the one-dimensional Hausdorff measure is in $\sum$), then (51) simply means that

(52)
$$\int_S \int_0^1 \beta(x,r)^2 \frac{d\mu(x)dr}{r} < +\infty.$$

Also, the condition for S to be contained in an Ahlfors-regular curve is then (48) (again under the condition that $S = \operatorname{supp}\mu$ for a $\mu \in \sum$).

In our study of singular integrals on a set S, the following very weak version of (48), or (50), is sometimes already useful. We shall say that " S satisfies a weak geometric lemma" if, for each $\epsilon > 0$, one can find a constant $C(\epsilon)$ such that, if $\mathcal{B} = \{(x,t) \in S \times \mathbb{R}_+ : \beta(x,r) > \epsilon\}$, then the measure $1_{\mathcal{B}}(x,r)d\mu(x)\frac{dr}{r}$ is a Carleson measure with norm $\le C(\epsilon)$. In other words,

(53)
$$\int\int_{[S\cap B(X,R)]\times]0,R]\cap\mathcal{B}} \frac{d\mu(x)dr}{r} \le CR^k$$

for all $X \in S$ and $R > 0$.

It is easily proved that Lipschitz graphs always satisfy (53). Also, sets in $S(k)$ satisfy a little more than (53). This is proved by Semmes in [Se8] ; amusingly, the proof comes from a variant of the square function estimate (46), applied to the unit normal $n(y)$, and so uses much more "operator theory" than geometrical arguments. We shall also see in a later subsection that if $S = \operatorname{supp} \mu$ for a $\mu \in \sum$, and if the square function estimates (45) hold, then S satisfies a weak geometric lemma.

Conversely, a weak geometric lemma is not enough by itself to imply that S is rectifiable. However, if $S = \operatorname{supp} \mu$ for some $\mu \in \sum$, if S satisfies a weak geometric lemma, and furthermore S "has big projections" (as defined in the beginning of Section 7), then S CBPLG (see Definition 3.4). The proof is a minor modification of P. Jones' proof of Theorem 8.1 given above (only Lemma 8.4 needs to be modified to work when f is a projection defined on S). Proving directly that all surfaces $S \in S(k)$ satisfy a weak geometric lemma is not too hard, and this can be used to give yet another proof of the fact that $S \in S(k)$ always CBPLG. See [DS4] for details.

C - The corona construction

If S satisfies the weak geometric lemma, then the methods of Carleson's celebrated construction provide a very powerful tool for analyzing the geometry of S and relating it to analytical issues. The idea for this comes from work of Garnett and Jones [GT] and Jones [Jn1], in which they use the corona construction to decompose complicated domains into simpler pieces, with an estimate on how the pieces fit together. Here, we intend to do the stopping time arguments directly on the set S (rather than through a conformal mapping, for instance). Thus it is a good idea to use an analogue of dyadic cubes on S.

Let us recall what we mean by this. We shall use a collection $\mathcal{R}$ of subsets $Q \subset S$, with the following properties. First, $\mathcal{R} = \bigcup_{j \in \mathbb{Z}} \mathcal{R}_j$, where $\mathcal{R}_j$ will be thought of as the set of dyadic cubes of size 2^{-j}. We want that

(54) for each $j \in \mathbb{Z}$, $S = \bigcup_{Q \in \mathcal{R}_j} Q$ is a partition of S ;

(55) if $j \le j'$, $Q \in \mathcal{R}_j$ and $Q' \in \mathcal{R}_{j'}$, then either Q contains Q', or else $Q \cap Q' = \emptyset$;

(56) if $j \in \mathbb{Z}$ and $Q \in \mathcal{R}_j$, then $C^{-1}2^{-j} \le \operatorname{diam} Q \le C2^{-j}$, and $C^{-1}2^{-kj} \le \mu(Q)$ $\le C2^{-kj}$;

(57) if $j \in \mathbb{Z}$, $Q \in \mathcal{R}_j$, and $0 < \tau < 1$, then

$$\mu\{x \in Q \ : \ \operatorname{dist}(x, S\backslash Q) \le \tau 2^{-j}\} \le C\tau^{1/C}2^{-kj}.$$

The condition (57) is a precise way of saying that the cubes of $\mathcal{R}$ have small boundaries. It is useful in some applications, but it is not really needed in what follows.

The existence of a collection of cubes $\mathcal{R}$ such that (54)-(57) hold was proved in [Dv6]. However, since the proof was unduely complicated, we shall give a simpler variant in the appendix. The existence of $\mathcal{R}$ was recently generalized by M. Christ to the case of any space of homogeneous type. Moreover, his proof is not significantly more complicated, so the reader can also consult [Ch2] directly.

If S satisfies a weak geometric lemma, it is possible, for each $\epsilon > 0$, to split $\mathcal{R}$ into $\mathcal{R} = \mathcal{B} \cup \mathcal{G}$, where $\mathcal{B}$ satisfies the Carleson measure condition

$$\sum_{\substack{Q \in \mathcal{B} \\ Q \subset R}} \mu(Q) \leq C(\epsilon)\mu(R) \text{ for each cube } R \in \mathcal{R}, \tag{58}$$

and

(59) for each $Q \in \mathcal{G}$, there is a k-plane P_Q such that $\mathrm{dist}(y, P_Q) \leq \epsilon \,\mathrm{diam}\, Q$ for all $y \in S$ such that $\mathrm{dist}(y, Q) \leq \mathrm{diam}\, Q$.

Given a small $\delta > \epsilon$, it is also possible to split further the set $\mathcal{G}$ into disjoint "stopping time regions" S_j, $j \in J$, in such a way that $\mathcal{G} = \bigcup_{j \in J} S_j$, but also

(60) each S_j has a maximal element, for inclusion, denoted by $Q(S_j)$ and, if $Q \in S_j$ and $Q' \in \mathcal{R}$ is such that $Q \subset Q' \subset Q(S_j)$, then $Q' \in S_j$, too ;

(61) if $Q \in S_j$, then $\mathrm{Angle}(P_Q, P_{Q(S_j)}) \leq \delta$;

(62) if Q is one of the minimal cubes of S_j (again for inclusion), then one of the sons of Q is in the bad set $\mathcal{B}$, or else $\mathrm{Angle}(P_Q, P_{Q(S_j)}) \geq \frac{\delta}{2}$

(if $Q \in \mathcal{R}_j$ is a cube of size 2^{-j}, the sons of Q are the cubes $Q' \in \mathcal{R}_{j+1}$ such that $Q' \subset Q$; note that in our case, Q might have only one son, but it always has $\leq C$ sons).

Up to now, if $\mu \in \sum$, $S = \mathrm{supp}\, \mu$ and S satisfies a weak geometric lemma, it is not hard to construct, for all $0 < \epsilon < \delta$ small enough, the sets $\mathcal{B}$, $\mathcal{G}$, and S_j, $j \in J$, in such a way that (58) - (62) be satisfied. We shall say that S "admits a corona decomposition" if, furthermore, it is possible to choose $\mathcal{B}$, $\mathcal{G}$, and the S_j's in such a way that the $Q(S_j)$'s satisfy the following Carleson measure packing condition :

$$\sum_{j \in J : Q(S_j) \subset R} \mu(Q(S_j)) \leq C(\epsilon, \delta)\mu(R) \tag{63}$$

for each cube $R \in \mathcal{R}$.

This new condition looks a little technical, but it is extremely useful. We shall try to say a few words to that extent in the next section, but let us mention now that a very similar construction was already used in [Se9] to give a new proof of the boundedness of the operator T_μ^* when K is any "good kernel" and $S \in S(k)$. The boundedness of T_μ^* is obtained from the "corona decomposition" above, which is itself obtained using square functions estimates such as (46). The fact that $S \in S(k)$ admits a corona decomposition also allows Semmes to prove the estimate on harmonic measure mentioned in Section 7 (see Corollary 7.3).

Let us finally mention that the first use of the corona construction in this context is due to P. Jones, in his study of Ahlfors-regular curves (see [Jn1]) ; however, the argument still used a conformal mapping, and the stopping time was performed on the unit disk.

D - A characterization

Theorem 9.5 [DS3]. *For a k-dimensional set $S \subset \mathbb{R}^n$ such that $S = \operatorname{supp} \mu$ for a $\mu \in \sum$, the following properties are equivalent :*

a) for each good kernel (see Definition 1.1), the operators T^ϵ_μ, $\epsilon > 0$, defined by (4) are uniformly bounded on $L^2(S, d\mu)$;

b) for each good kernel K, the operator T^*_μ defined by (5) is bounded on $L^p(S, d\mu)$ for $1 < p < +\infty$;

c) the square function estimate (45) holds for every odd function $\psi \in C^\infty_c(\mathbb{R}^n)$;

d) S admits a corona decomposition (i.e., one can find decomposition of the set $\mathcal{R}$ of dyadic cubes on S that satisfy (58) - (63)) ;

e) the measure $\beta_1(x,r)^2 \frac{d\mu(x)dr}{r}$ is a Carleson measure on $S \times \mathbb{R}^+$ (see (49) for a definition of β_1) ;

f) the measure $\beta_p(x,r)^2 \frac{d\mu(x)dr}{r}$ is a Carleson measure for $1 \leq p < \frac{2k}{k-2}$ when $k > 1$, and for $1 \leq p \leq +\infty$ when $k = 1$ (in other words, (50) is satisfied) ;

g) letting $n_1 = Max(n, 2k+1)$, for each $\epsilon > 0$ there exists a constant $M \geq 1$ and, for all $x \in S$ and $r > 0$, an M-bilipschitz mapping $z : \mathbb{R}^k \to \mathbb{R}^{n_1}$ such that $\mu(S \cap B(x,r) \backslash z(\mathbb{R}^k)) \leq \epsilon r^k$ (if $n_1 > n$, $\mathbb{R}^n$ is embedded in $\mathbb{R}^{n_1}$ in the natural way) ;

h) letting $n_2 = n$ if $n \geq 2k$ and $n_2 = n+1$ otherwise, there is an ω-regular function $z : \mathbb{R}^k \to \mathbb{R}^{n_2}$ (see Definition 6.4) such that $S \subset z(\mathbb{R}^k)$.

Let us first comment rapidly on the proofs. The equivalence of a) and b) is standard Calderón-Zygmund theory. The fact that a) implies c) is easy, and the fact that g), or h) implies b) is easily deduced from the results of previous sections (see Proposition 3.2, Corollary 3.6 or Theorem 6.5, for example). What really helps proving the theorem, however, is the condition d). Indeed, if S_j is one of the stopping-time regions with the properties (59), (60) and (61), then one can construct a Lipschitz graph Γ_j, which is a very good approximation of S, at the scale of all the cubes of S_j. To prove that d) implies f), g), or h), one uses the fact that f), g) and h) are satisfied by the Lipschitz graphs Γ_j, and one tries to glue together the various pieces coming from the various S_j's. The inequalities

(58) and (63) precisely control the amount of gluing that has to be done, and altogether deducing f), g) and h) from d) is relatively easy.

The hard part of the theorem is deducing d) from c) and, to a lesser extent, from e). What saves us in this case is that it is not necessary to have d) to construct the S_j's, and then the lipschitz graphs Γ_j : only a weak geometric lemma is needed there. The technical part of the proof is moving square function estimates from S onto the lipschitz graphs Γ_j ; the idea of the proof is that, because of the second part of (62), the square function estimate on each Γ_j cannot be too good (one does not expect Γ_j to be much better than a δ-lipschitz graph). One then controls the number of stopping-time regions S_j by associating, to each of them, a piece of square function which is not too small ; (63) then follow from (45). A few more details can be found in [DS3].

Let us conclude this section with a few comments on Theorem 9.5.

In condition h), one can take $\omega \in A_1(\mathbb{R}^k)$. We do not know, however, if one can take $\omega = 1$.

The conditions e) and f) are not that much different from each other ; they can be seen as quantified ways of saying that S is "rectifiable", or "regular" in the sense of Besicovitch (see [Fe], [Mt], or [Fa] for definitions). Actually, conditions like "S CBPLG", or g), or even h), can be seen as stronger rectifiability properties, too (compare them to the property of being contained in the union of a countable family of C^1-surfaces, which is one of the definitions of rectifiability).

Also, the fact that e) implies h) can be seen as an extension of P. Jones traveling salesman theorem (Theorem 9.4 below) : we characterize the Ahlfors-regular sets S that are contained in the image of an ω-regular function. Note that, when $k = 1$ and $n = 2$, our result is weaker than Jones', because we restrict ourselves to Ahlfors-regular sets. On the other hand, we can use the slightly different functions β_p, $1 \leq p \leq +\infty$, instead of just β_∞.

The numbers n_1 of g) and n_2 of h) were introduced for technical reasons ; we do not know if they are sharp, or even if it is impossible to take $n_1 = n_2 = n$ (provided that $k < n$).

The reader might find it interesting to consult [DS5] or [Dv8], where the theorem is presented with a slightly different point of view.

Finally, note that we did not say anything about the case when k is not an integer; it is actually part of the theorem that a), b) or c) is never satisfied in that case. This of course does not mean that no interesting kernel will ever define a bounded operator on a snowflake, for instance, but we should not expect all "good kernels" to define bounded operators at the same time.

10. A few questions

Let us start with questions relative to parametrizations. When $k = 2$ (and $n = 3$), S. Semmes proved that every CASSC (see Section 6.A) is the image of $\mathbb{R}^2$ by a quasisymmetric mapping and so, in particular, can be parametrized in an ω-regular way (see Section 6-B).

His proof relies on the existence of a conformal mapping from $\mathbb{R}^2$ to S, and does not extend to higher dimensions (see [Se4]).

Is it possible to parametrize all CASSC's of dimension $k = 2$ by 1-regular mappings? Stated differently, this is a question on the class I_0 of weights $\omega \in A_\infty$ that show up as the Jacobian of a quasiconformal mapping of the plane : is it true that if ω is the Jacobian of a conformal mapping from $\mathbb{R}^2$ onto a CASSC S, then $\omega = b\omega_0$ for some $\omega_0 \in I_0$ and some b such that $\log b \in L^\infty$? If so, a quasiconformal change of variable in $\mathbb{R}^2$ gives a 1-regular parametrization of S from any conformal parametrization.

One can also try to characterize the weights $\omega \in I_0$, or, less ambitiously, their classes modulo multiplication by the exponential of a bounded function. Obviously, not all A_∞−weights are good (exercise : if ω is continuous, and its restriction to a C^∞ curve is zero, then ω cannot be in I_0). In fact, if ω is the Jacobian of a quasiconformal mapping of $\mathbb{R}^k$, $k \geq 2$, or even if $\omega \in A_\infty(\mathbb{R}^k)$ is such that there exists an ω-regular mapping, then ω has to satisfy the following "strong-A_∞ condition" : there is a constant $C > 0$ such that, if $B \subset \mathbb{R}^k$ is a ball and γ is a path that joins the center of B to its boundary, then

$$\int_\gamma \omega^{1/k} \geq C^{-1}\left\{\int_B \omega\right\}^{1/k} . \tag{64}$$

We do not know, however, if every strong-A_∞ weight(or even every strong-A_∞ weight with $\| \log \omega \|_{BMO}$ small enough) is such that there exists an ω-regular mapping. For more details on strong-A_∞ weights, see [DS2], and for a partial result in this direction, see [Se11].

It seems reasonable to conjecture that every CASSC of dimension $k > 2$ admits an ω-regular parametrization (or even a quasisymmetric one). Reasonable parametrizations were found by Semmes [Se4], but they do not quite have the right scale invariance ; also, note that part h) of Theorem 9.5 only gives an ω-regular parametrization of a set which is larger than S.

One can try to relate Sobolev estimates on S to the boundedness of singular integrals on $L^2(S)$. Let us give an example of Sobolev estimates. Suppose that $S \in S(k)$, and make all the necessary a priori assumptions. Define

$$Nf(x) = \sup_{r>0} r^{-1}\left\{r^{-k}\int_{S\cap B(x,r)} | f(y) - f(x) |^2 \, d\mu(y)\right\}^{1/2}$$

and

$$osc_a f(x,r) = \inf_A \left\{r^{-k}\int_{S\cap B(x,r)} | f(y) - A(y) |^2 \, d\mu(y)\right\}^{1/2} ,$$

where f is, say, of class C^1 on S and the infimum is taken over all Clifford-analytic, affine functions A. Semmes proves that

$$\int_S Nf(x)^2 d\mu(x) \sim \int_0^\infty \int_S [t^{-1} osc_a f(x,t)]^2 \frac{d\mu(x)dt}{t}. \tag{65}$$

Furthermore, if each of the two connected components of $\mathbb{R}^{k+1} \setminus S$ is a NTA domain, then the two quantities above are equivalent to $\int_S |df|^2$, where df denotes the differential of f on S. In the general case, S might have points where it chokes, and then $|f(x) - f(y)|$ could be large even if x and y are close and $|df|$ is small (think about a curve which is not chord-arc), and we cannot expect as much. It is still possible to give an estimate, by controlling the number of choking points of S. We refer to [Se8] for more details.

It is not clear whether we should expect the right Sobolev estimates to imply that singular integrals are bounded. This looks like a reasonable way to replace the assumption of connectedness (or local connectedness) which implies the boundedness of singular integrals when S is Ahlfors-regular and one-dimensional. We know that, when $k \geq 2$, local connectedness is not enough : see [Dv6, p. 113] for the picture of an example. It would probably be interesting to study a little more the theory of differentiability of functions on one of our preferred surfaces S (for instance, is there anything like Littlewood-Paley theory on S ?). Note that the corona decomposition already gives some information.

Theorem 9.5 does not tell us everything we wanted to know about singular integrals on surfaces. The most important problem is that conditions a) and c) use much more kernels and functions ψ than we would like. For instance, if $k = 1$ and $n = 2$, we would really like to characterize the Ahlfors-regular sets S such that the Cauchy kernel (alone !) defines a bounded operator on $L^2(S)$. We do not even know if it is enough to consider, in a), all the odd kernels K that are homogeneous of degree $-k$. When $k = 1$ and $n = 2$, knowing whether the boundedness of T_μ^* when K is the Cauchy kernel implies the boundedness of all other T_μ^*'s would have nice applications to the theory of analytic capacity.

Even if we decide to consider all good kernels, the equivalent conditions of Theorem 9.5 are not necessarily easy to check. In fact, it seems that in all known examples, it is just as simple to prove directly that S CBPLG ! Here is an example of a class of surfaces, for which we do not know whether Theorem 9.5 applies. Suppose S is a k-dimensional surface, homeomorphic to $\mathbb{R}^k$, contained in $\mathbb{R}^{k+1}$, and such that $S = \operatorname{supp} \mu$ for some $\mu \in \sum$. Also suppose that for each $x \in S$ and $r > 0$, there is a topological ball of dimension k, contained in $S \cap B(x, r)$, but which contains $S \cap B(x, \frac{r}{c})$. What can be said about S ? When $k = 2$, Semmes used a conformal mapping and modulus estimates to show that S is the image of $\mathbb{R}^2$ by a quasisymmetric mapping (and so, S CBPLG), but in higher dimensions nothing is very clear.

Some consequences of Theorem 9.5 have, at this moment, surprisingly indirect proofs. Let us give a few examples to amuse the reader. To prove that a), or b), is invariant when S is replaced by its image under a bilipschitz mapping, it seems that we need the equivalence with one of the last five conditions of the theorem. Similarly, if we want to know that, if S "contains big pieces of sets that contain big pieces .... of Lipschitz graphs", then S contains big pieces of sets that contain big pieces of Lipschitz graphs, or even d), e), f), g) or h), we seem to have to go through singular integral operators ! Finally, Proposition 3.8 tells us that, if the equivalent conditions of Theorem 9.5 are satisfied, then $\lim_{\epsilon \to 0} T_\mu^\epsilon f(x)$ exists almost everywhere for every good kernel K. It is not clear how to prove this directly (and in particular under the weaker assumption that T_μ^* is bounded for a single K).

We still do not know whether the equivalent properties of Theorem 9.5 also imply that S CBPLG. Since S satisfies a weak geometric lemma, this is equivalent to asking whether

S has big projections. Using condition g), we see that it would be enough to prove the following. If S is the image of $\mathbb{R}^k$ by a bilipschitz mapping, and if $\theta > 0$ is given, there is a constant $\nu > 0$ (that depends only on k, n, the bilipschitz constant and θ) such that, whenever $x \in S$, $r > 0$, and $E \subset S \cap B(x,r)$ is a compact set satisfying $\mu(E) \geq \theta r^k$, there is a k-plane V such that $| \pi_v(E) | \geq \nu r^k$ (π_v denotes the orthogonal projection on V). Amusingly enough, this does not even seem to be known (or known to be wrong) when $k = 1$ and $n = 2$!

Let us finally quote a variant of Vituschkin's conjecture on sets of vanishing analytic capacity. Let us restrict to $k = 1$ and $n = 2$ (the question is probably hard enough in that case !). Suppose $\mu \in \sum$, $S = \text{supp } \mu$, and $0 \in S$, and consider $S_0 = S \cap B(0,1)$. The Favard length of S_0 is the number $FL(S_0) = \int_0^\pi | \pi_\theta(S_0) | \, d\theta$, where, for each $\theta \in [0,\pi]$, π_θ is the orthogonal projection on a line that makes the angle θ with the real axis.

Is it true that if $FL(S_0) \geq a > 0$, then there is an image by a rotation of a Lipschitz graph, Γ, such that $\mu(\Gamma \cap S_0) \geq \nu > 0$? We would like ν, and the Lipschitz constant of Γ, to depend only on a and the Ahlfors-regularity constant of μ. Would it help if we supposed that the Favard length is large at all scales ? What about supposing that $| \pi_\theta(S_0) | \geq a > 0$ for all θ in a small interval ? Of course, we are interested in this question because the existence of Γ would imply that the analytic capacity of S is > 0.

If these questions are not hard enough, one can always try to say something sensible in the case when $\mu \in \Delta$, but μ is not necessarily in $\sum$. For instance, does a variant of the implication $e) \Rightarrow h)$ (the traveling salesman theorem in higher dimensions) hold when one does not assume that $S = \text{supp } \mu$ for a $\mu \in \sum$?

I hope, too, that at this stage, the reader will be tempted to add a few arrows of his own (or, better, a few boxes) to the diagrams of Appendix II.

APPENDIX I

Construction of dyadic cubes on a regular set

Let $\mu \in \sum$ (see Definitions III.2.3 and III.1.3), and let S be the support of $\sum$. We want to construct a family $\mathcal{R}$ of "dyadic dubes" with the properties (54)-(57) described in Section III.9. The proof that follows is a minor modification of the argument in [Dv6] ; for an extension to spaces of homogeneous type (or a different proof), see [Ch 2].

Let us start by constructing balls with reasonably small boundaries. Let $\eta > 0$ be a small constant (to be chosen later).

Lemma A.1. *For each $x \in S$ and $j \in \mathbb{Z}$, there is a ball $B_j(x)$ centered at x and with a radius $r \in (2^{-j}, (1+\eta)2^{-j})$, and such that*

$$\begin{aligned}(A1)\qquad &\mu(\{y \in B_j(x) \ :\ \mathrm{dist}(y, S\backslash B_j(x)) < \eta^2 2^{-j}\})\\ &+\mu(\{y \in S\backslash B_j(x) \ :\ \mathrm{dist}(y, B_j(x)) < \eta^2 2^{-j}\}) \leq C\eta 2^{-kj}.\end{aligned}$$

This is fairly easy : one can find more than $\frac{1}{C\eta}$ radii r such that the sets that arise in (A1) are all disjoint. Since all these sets are contained in $B(x, 2^{-j+1})$, their total mass is $\leq C2^{-kj}$, and (A1) follows by the piegonhole principle.

Let us now construct coverings of S. For each $j \in \mathbb{Z}$, let $A_0(j)$ be a maximal set of points $x \in S$ with the property that $| x' - x | \geq 2^{-j}$ for $x \neq x' \in A_0(j)$. By the maximality of $A_0(j)$, the balls $B_j(x)$, $x \in A_0(j)$, are a covering of S. Also, since the balls $B(x, 2^{-j-1})$ are disjoint and have a mass $\geq \gamma 2^{-kj}2^{-k}$, there are never more than C points $y \in A_0(j)$ such that $B_j(y)$ meets $B(x, 2^{-j+1})$.

Our coverings are not partitions yet. To fix this, we put an order (denoted by $<$) on each $A_0(j)$, and we replace $B_j(x)$ by

$$(A2)\qquad B'_j(x) = B_j(x) \cap \left\{ \bigcup_{y<x} B_j(y) \right\}^c$$

(by what we just said, the union is locally finite).

The $S \cap B'_j(x)$, $x \in A_0(j)$, are now a partition of S, but we could have removed so much from $B_j(x)$ that $B'_j(x)$ will no longer satisfy (56). So we'll now glue some of the $B'_j(x)$'s together.

To do so, let $A(j)$ be the set of $x \in A_0(j)$ such that $\mu(B'_j(x)) \geq C_0^{-1}2^{-kj}$, where the constant C_0 will be chosen soon. To each $x \in A_0(j)\backslash A(j)$, let us associate a point $h(x) \in A(j)$ as follows. Note that there are less than C points $y \in A_0(j)$ such that $B'_j(y)$ meets $B(x, 2^{-j})$, and so one of them must be such that $\mu(B(x, 2^{-j}) \cap B'_j(y)) \geq C^{-1}\gamma 2^{-jk}$ (this is because the $B'_j(y) \cap S$ are a partition of S). We let $h(x)$ be such a y, and we choose $C_0 > C\gamma^{-1}$, so that $h(x) \in A(j)$. Note that $| h(x) - x | \leq 2^{-j+2}$.

For each $y \in A(j)$, we set

$$(A3)\qquad D_j(y) = S \cap \left\{ B'_j(y) \cup [\bigcup_{h(x)=y} B'_j(x)] \right\}.$$

The $D_j(y)$, $y \in A(j)$, are a new partition of S, and they satisfy

$$\text{(A4)} \qquad \operatorname{diam} D_j(y) \le 2^{-j+3},$$

$$\text{(A5)} \qquad \mu(D_j(y)) \ge C_0^{-1} 2^{-jk},$$

and

$$\text{(A6)} \qquad \begin{aligned} &\mu(\{u \in D_j(y) \ : \ \operatorname{dist}(u, S \backslash D_j(y)) < \eta^2 2^{-j}\}) \\ &+\mu(\{u \in S \backslash D_j(y) \ : \ \operatorname{dist}(u, D_j(y)) < \eta^2 2^{-j}\}) \le C\eta 2^{-kj}. \end{aligned}$$

Let us try now to make the various generations of cubes interact nicely with each other. To do so, it will be easier to skip some of the generations. Let N be a large integer (it will be chosen soon, so that 2^{-N} is smaller than η^2) ; we shall restrict ourselves to the j's that are multiples of N.

If $j \in N\mathbb{Z}$ and $x \in A(j)$, let $\varphi(x)$ be the point $y \in A(j-N)$ such that $x \in D_{j-N}(y)$. Next, for $j \in N\mathbb{Z}$ and $d \in \mathbb{N}$, let $E_d(x) = \{y \in A(j+Nd) \ : \ \varphi^d(y) = x\}$, where $\varphi^d = \varphi \circ \cdots \circ \varphi$. We now replace $D_j(x)$ by unions of smaller cubes :

$$\text{(A7)} \qquad D_{j,d}(x) = \bigcup_{y \in E_d(x)} D_{j+Nd}(y).$$

Note that the Hausdorff distance between the sets $D_j(x)$ and $D_{j,1}(x)$ is trivially $\le C2^{-j}$. Then, since $D_{j,d+1}(x) = \bigcup_{y \in E_d(x)} [D_{j+Nd,1}(y)]$, the Hausdorff distance between the sets $D_{j,d}(x)$ and $D_{j,d+1}(x)$ is less than $C2^{-j-Nd}$. Consequently, for $j \in N\mathbb{Z}$ and $x \in A(j)$ fixed, the sequence of the closures $\overline{D_{j,d}(x)}$ converges, for the Hausdorff topology, to a compact set $R_j(x)$.

First note that the $R_j(x)$, $x \in A(j)$, are still a covering of S. Indeed, fix $u \in S$ and j ; for each $d \ge 0$, there is an $x \in A(j)$ such that $u \in D_{j,d}(x)$. Since there is only a finite set of points x that can arise this way, u is in infinitely many $D_{j,d}(x)$ for some x, and so $u \in R_j(x)$. However, because we took closures, the $R_j(x)$'s are no longer a partition of S. We could content ourselves with them and show the intersections have measure 0, but let us go ahead and modify the $R_j(x)$'s again.

For each $j \ge 0$, put an order (denoted by $<$) on $A(j)$, and do so in such a way that $x < y$ as soon as $\varphi(x) < \varphi(y)$. For all $j \ge 0$ and $x \in A(j)$, set

$$\text{(A8)} \qquad Q_j(x) = R_j(x) \cap \left(\bigcup_{\substack{y \in A(j) \\ y < x}} R_j(y) \right)^c .$$

Note that, as before, the union is locally finite. Obviously, for each $j \ge 0$, the $Q_j(x)$, $x \in A(j)$, are a partition of S. Let us check that they also satisfy the conditions (55), (56) and (57) of Part III.9.

To do so, we first check how close $Q_j(x)$ is to $D_j(x)$. Let $j \in N\mathbb{Z}$, $x \in A(j)$ and $u \in D_{j,1}(x)$. If y is the point of $A(j+N)$ such that $u \in D_{j+N}(y)$, then $y \in D_j(x)$ and so $\mathrm{dist}(u, D_j(x)) \le 2^{-j-N+1}$. We have seen before that the Hausdorff distance between $D_{j,1}$ and any $D_{j,d}$, $d > 1$, is $\le C2^{-j-N}$, and so $\mathrm{dist}(u, D_j(x)) \le C2^{-j-N}$ for every $u \in R_j(x)$.

Next, if $u \in D_j(x) \cap Q_j(x)^c$, then $u \in R_j(x')$ for some $x' \neq x$ and consequently $\mathrm{dist}(u, S \backslash D_j(x)) \le C2^{-j-N}$. We choose N so that $2^{-j-N} < \eta^2 2^{-j-2}$. It follows from (A6) that the symmetric difference between $D_j(x)$ and $Q_j(x)$ has measure $\le C\eta 2^{-kj}$.

Using this and (A5), we get that $\mu(Q_j(x)) \ge C_0^{-1} 2^{-kj-1}$ (provided η is small enough). Since $\mathrm{diam}\, Q_j(x) \le 2^{-j+4}$ and $\mu \in \sum$, we see that $Q_j(x)$ satisfies (III.56). Moreover, let us check that

$$(A9)\qquad \begin{aligned} &\mu(\{u \in Q_j(x) \ :\ \mathrm{dist}(u, S \backslash Q_j(x)) \le \eta^2 2^{-j-1}\}) \\ &+\mu(\{u \in S \backslash Q_j(x) : \mathrm{dist}(u, Q_j(x)) \le \eta^2 2^{-j-1}\}) \le C\eta 2^{-kj}. \end{aligned}$$

Indeed, for the first set, note that if $u \in D_j(x)$ and $\mathrm{dist}(u, S \backslash Q_j(x)) \le \eta^2 2^{-j-1}$, then $\mathrm{dist}(u, D_j(x')) \le \eta^2 2^{-j}$ for some $x' \neq x$. The corresponding set has small measure by (A6). Since $Q_j(x) \cap D_j(x)^c$ is also small, we get the desired estimate for the first set. The second set is a finite union of sets like the first one ; (A9) follows.

We did not define $Q_j(x)$ when $j < 0$ yet. Let us do it now : if $j = -Nd$ for some $d > 0$ and $x \in A(j)$, let

$$(A10)\qquad Q_j(x) = \bigcup_{\substack{y \in A(0) \\ \varphi^d(y) = x}} Q_0(y).$$

For the same reasons as above, the $Q_j(x)$, $x \in A(j)$, are partitions of S, satisfy (III.55), (III.56) and (A9). We shall now check that the more precise property (III.57) follows from this.

Fix $j \in N\mathbb{Z}$, $x \in A(j)$ and $\tau > 0$, and let

$$F = \{u \in Q_j(x) \ :\ \mathrm{dist}(u, S \backslash Q_j(x)) \le \tau 2^{-j}\}$$

$$\cup \{u \in S \backslash Q_j(x) \ :\ \mathrm{dist}(u, Q_j(x)) \le \tau 2^{-j}\}.$$

For every $d \ge 0$, let $B(d)$ be the set of all $y \in A(i + Nd)$ such that $Q_{j+Nd}(y)$ meets F, and let $F_d = \bigcup_{B(d)} Q_{j+Nd}(y)$. Obviously, $F_0 \supset F_1 \cdots \supset F_d \supset F$.

Now suppose that $\tau < 2^{-N(d+1)}$ and let $y \in B(d)$. If $u \in F_{d+1} \cap Q_{j+Nd}(y)$, then $\mathrm{dist}(u, F) \le 2^{4-j} 2^{-N(d+1)}$ and so $\mathrm{dist}(u, S \backslash Q_{j+Nd}(y))$ is less than $\mathrm{dist}(u, S \backslash Q_j(x))$ if $u \in Q_j(x)$ and than $\mathrm{dist}(u, Q_j(x))$ otherwise. Consequently, $\mathrm{dist}(u, S \backslash Q_{j+Nd}(y)) \le 2^{4-j} 2^{-N(d+1)} + \tau 2^{-j} < \eta^2 2^{-Nd} 2^{-j}$ by our choice of N. By (A9), $\mu(F_{d+1} \cap Q_{j+Nd}(y)) \le \frac{1}{10} \mu(Q_{j+Nd}(y))$, provided we chose η much smaller than C_0^{-1}. Summing on $y_0 \in B(d)$ gives $\mu(F_{d+1}) \le \frac{1}{10} \mu(F_d)$, and repeating the argument gives $\mu(F_{d+1}) \le 10^{-d-1} \mu(F_0) \le C\, 10^{-d} 2^{-kj}$. This is true as long as $\tau < 2^{-N(d+1)}$, and so we get $\mu(F) \le C\tau^{1/C} 2^{-kj}$, as needed.

We now take for $\mathcal{R}_j$ the set of $Q_j(x)$, $x \in A(j)$ when $j \in N\mathbb{Z}$, and $\mathcal{R}_{aN+b} = \mathcal{R}_{aN}$ whenever $0 < b < N$; we just proved that the R_j's satisfy the conditions (54) - (57) of III.9.

Remark. In our construction, many of the cubes have only one child (themselves). This does not create any problem in the arguments, and cannot be avoided if you want to use the dyadic scheme and the dimension k is less than 1, say.

APPENDIX II

Two recapitulatory diagrams

In the following two pictures, we used the following abbreviations :

- Bilipschitz images are the images of $\mathbb{R}^k$ under a bilipschitz mapping $z : \mathbb{R}^k \to \mathbb{R}^n$;

- CASSC's are Chord-arc surfaces with small constant, as in Section 6.A ;

- Sobolev inequalities are alluded to just after (65) ;

- $\mathcal{C}_S$ is the Cauchy-Clifford operator, as in Sections 6.A and 6.C ;

- "T^*_μ is bounded" means that for every good kernel K (see Definition 1.1), the operator T^*_μ defined by (4) and (5) is bounded on $L^2(S, d\mu)$;

- Semmes surfaces are the sets $S \in \mathcal{S}(k)$ introduced in Section 6.C ;

- W.G.L. stands for "weak geometric lemma". It is the weak version of P. Jones' geometric lemma introduced in Section 9.B, (53) ;

- "Projections" refers to the property "S has big projections", as in the beginning of Section 7 ;

- "S $CBPLG$" means "S contains big pieces of Lipschitz graphs", as in Definition 3.4;

- Harmonic measure estimates are estimates that say that, if $\mathcal{O}$ is one of the components of $\mathbb{R}^{k+1} \backslash S$, and if $\mathcal{O}$ is a non-tangential access domain, then harmonic measure relative to $\mathcal{O}$ is A_∞-equivalent to surface measure on $\partial\mathcal{O}$ (see Corollary 7.3) ;

- Big disks surfaces are the following slight modification of the class $\mathcal{S}(k)$. In the condition (37), one replaces "$\mathrm{dist}(x_i, S) \geq \frac{r}{C}$" by "$x_i$ is the center of a k-dimensional disk of radius $\frac{r}{C}$ (or even the image of such a disk by a bilipschitz mapping) that does not meet S" ;

- "Higher codimension variant" refers to a generalization of the class $\mathcal{S}(k)$ to higher codimension ; we refer to [Dv6] for more details ;

- S(CBP)^{m}LG is, as in Remark 3.7, a generalization of the condition S CBPLG ;

- S CVBPBI (S contains very big pieces of bilipschitz images of $\mathbb{R}^k$) is the condition g) of Theorem 9.5 ;

- "contained in ω-regular" is the condition h) of Theorem 9.5 : S is contained in the image of $\mathbb{R}^k$ by some ω-regular mapping (with perhaps values in $\mathbb{R}^{n+1}$) ;

- "square function estimates" are the estimates (45) (condition c) of Theorem 9.5) ;

- the corona decomposition is the condition mentionned in Section 9.B (condition d) of Theorem 9.5) ;

- the Geometric Lemma is the variant of P. Jones' condition mentioned in conditions e) and f) of Theorem 9.5 ;

- ω-regular surfaces are the images of $\mathbb{R}^k$ by ω-regular mappings, as in Definition 6.4.

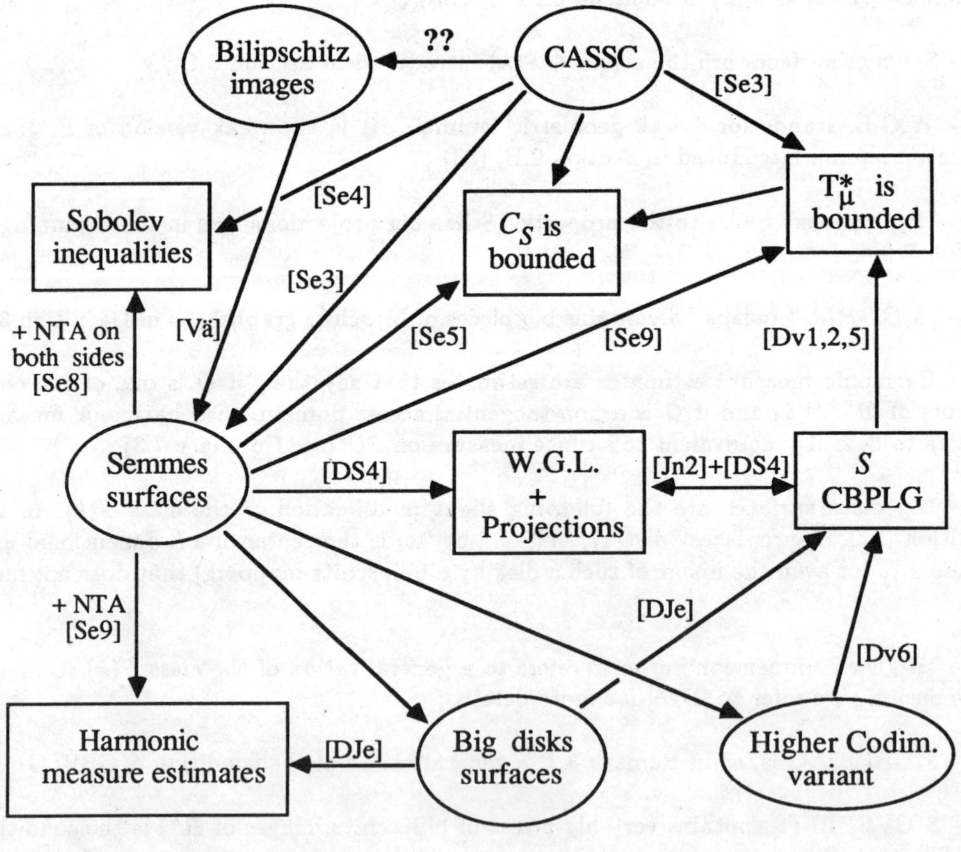

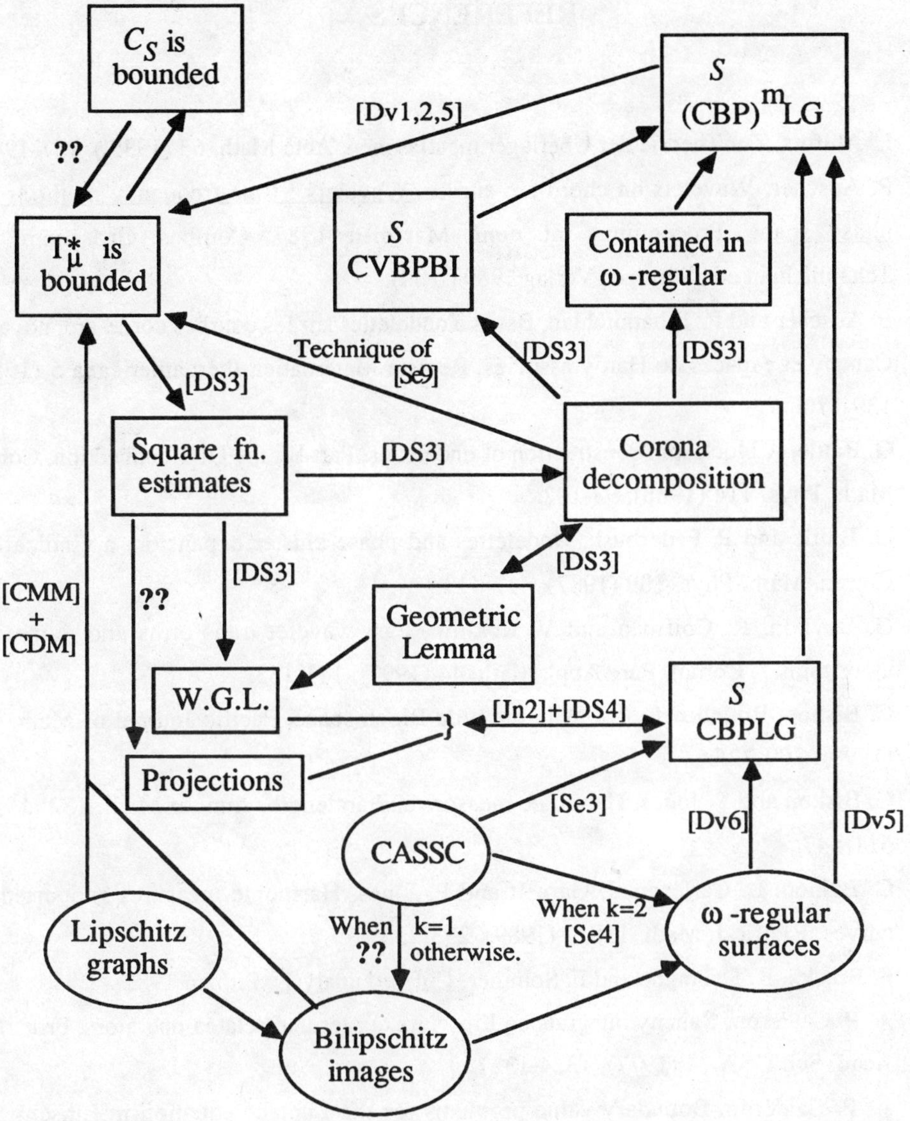

These two diagrams show some of the implications between properties of a set $\mathcal{S}$ which is the support of a measure μ in Σ.

REFERENCES

[Ah] L. Ahlfors, Zur Theorie der Uberlagerungsfläschen, Acta Math. 65 (1935), 157-194.

[Au] P. Auscher, Wavelets on chord-arc curves, Wavelets : time-frequency methods and phase space, Proceedings int. conf. Marseille 1987, Combes, Grossman and Tchamitchian ed., Springer-Verlag 1989 (IPTI).

[AT] P. Auscher and P. Tchamitchian, Bases d'ondelettes sur les courbes corde-arc, noyau de Cauchy et espaces de Hardy associés, Revista Matematica Iberoamericana 5 (1989), 139-170.

[Ba] G. Battle, A block spin construction of ondelettes, Part II : the QFT connection, Comm. Math. Phys. 114 (1988), 93-102.

[BF] G. Battle and P. Federbush, Ondelettes and phase cluster expansion, a vindication, Comm. Math. Phys. 109 (1987), 417-419.

[BCR] G. Beylkin, R. Coifman and V. Rokhlin, Fast wavelet transforms and numerical algorithms 1, Comm. Pure Appl. Math. 44 (1991), 141-183.

[Bi] C. Bishop, Bounded functions in the little Bloch space, Pacific Journal of Math. 142 (1990), 209-225.

[BJ] C. Bishop and P. Jones, Harmonic measure and arclength, Ann. of Math. 132 (1990), 511-547.

[BCJ] C. Bishop, L. Carleson, J. Garnett and P. Jones, Harmonic measures supported on curves, Pacific J. Math. 138, 2 (1989), 233-236.

[BDS] F. Brackx, R. Delanghe and F. Sommer, Clifford analysis, Pitman 1982.

[Ca1] A. P. Calderón, Cauchy integrals on Lipschitz curves and related operators, Proc. Nat. Acad. Sci. USA 74 (1977), 1324-1327.

[Ca2] A. P. Calderón, Boundary value problems for the Laplace equation in Lipschitzian domains, Recent progress in harmonic analysis (El Escorial 1983), Ed. I. Peral and J-L. Rubio de Francia, North-Holland Mathematics Studies 111 (1985), 33-49.

[CZ] A. P. Calderón and A. Zygmund, On the existence of certain singular integrals, Acta Math. 88 (1952), 85-139.

[Ch1] M. Christ, Lectures on Singular integral operators, NSF-CBMS Regional Conference series in Mathematics 77, AMS 1990.

[Ch2] M. Christ, A T(b) theorem with remarks on analytic capacity and the Cauchy integral, preprint.

[ChJ] M. Christ and J-L. Journé, Polynomial growth estimates for multilinear singular integral operators, Acta Math. 159 (1987), 51-80.

[Co] R. Coifman, Adapted multiresolution analysis, computation, signal processing and operator theory, ICM 1990, Kyoto, Springer-Verlag.

[CJS] R. R. Coifman, P. Jones and S. Semmes, two elementary proofs of the L^2-boundedness of Cauchy integrals on Lipschitz curves, Journ. A.M.S. vol. 2, 3 (1989), 553-564.

[CDM] R. R. Coifman, G. David and Y. Meyer, La solution des conjectures de Calderón, Advances in Math. 48 (1983), 144-148.

[CMM] R. R. Coifman, A. McIntosh and Y. Meyer, L'intégrale de Cauchy définit un opérateur borné sur L^2 pour les courbes lipschitziennes, Ann. of Math. 116 (1982), 361-387.

[CM1] R. R. Coifman and Y. Meyer, Au-delà des opérateurs pseudo-différentiels, Astérisque 57, Société Mathématique de France, Paris 1978.

[CM2] R. R. Coifman and Y. Meyer, Une généralisation du théorème de Calderón sur l'intégrale de Cauchy, Proceedings of the seminar on Fourier analysis, El Escorial 1979, edited by M. De Guzman and I. Peral, Asociation Matematica Española 1980.

[CM3] R. R. Coifman and Y. Meyer, Orthonormal wave packet bases, preprint.

[CM4] R. R. Coifman and Y. Meyer, Remarques sur l'analyse de Fourier à fenêtre, C. R. Acad. Sc. Paris, t. 312, Série I (1991), 259-261.

[CMW1] R. R. Coifman, Y. Meyer and V. Wickerhauser, Size properties of wavelets packets, Wavelets, ed. by G. Beylkin etc., Jones & Bartlett, Boston.

[CMW2] R. R. Coifman, Y. Meyer and V. Wickerhauser, Adapted wave form analysis, wavelet packets and applications, to appear in Wavelets : Mathematics and Applications, Benedetto and Frazier ed., CRC Press.

[CS] R. R. Coifman and S. Semmes, Real-analytic operator-valued functions defined on BMO, Analysis and partial differential equations : a collection of papers dedicated to Mischa Cotlar, ed. C. Sadosky, Lecture notes in pure and applied mathematics 122, Marcel Dekker (1989), 85-100.

[CW] R. R. Coifman and G. Weiss, Analyse harmonique non-commutative sur certains espaces homogènes, Lecture Notes in Math. 242, Springer-Verlag 1971.

[CWi] R. R. Coifman and M. V. Wickerhauser, Best-adapted wave packet bases, preprint.

[Db1] I. Daubechies, Orthonormal bases of compactly supported wavelets, Comm. in Pure and Applied Math. 61, 7 (1988), 909-996.

[Db2] I. Daubechies, The wavelet transform, time-frequency localization and signal analysis, IEEE, Information Theory 36, 5 (1990), 961-1005.

[DbGM] I. Daubechies, A. Grossman and Y. Meyer, Painless non-orthogonal expansions, J. Math. Phys., vol. 27 (1986), 1271-1283.

[DbP] I. Daubechies and T. Paul, Wavelets and applications, Proceedings of the VIII-th International Congress of Mathematical Physics, M.Mebkhout and R.Seneor editors, World Scientific Publishers, 1987.

[Dv0] G. David, Courbes corde-arc et espaces de Hardy généralisés, Ann. Inst. Fourier Grenoble 3 (1982), 227-239.

[Dv1] G. David, Opérateurs intégraux singuliers sur certaines courbes du plan complexe, Ann. Sci. Ec. Norm. Sup. 17 (1984), 157-189.

[Dv2] G. David, Noyau de Cauchy et opérateurs de Calderón-Zygmund, Thèse d'état , Paris XI - Orsay 3193 (1986).

[Dv3] G. David, Une minoration de la norme de l'opérateur de Cauchy sur les graphes lipschitziens, Transactions A. M. S., Vol. 302, 2 (1987), 741-750 .

[Dv4] G. David, Opérateurs de Calderón-Zygmund, Proceedings I. C. M. Berkeley 1986, 890-899.

[Dv5] G. David, Opérateurs d'intégrale singulière sur les surfaces régulières, Ann. Sci. Ec. Norm. Sup., série 4, t.21 (1988), 225-258.

[Dv6] G. David, Morceaux de graphes lipschitziens et intégrales singulières sur une surface, Revista Matematica Iberoamericana, vol. 4, 1 (1988), 73-114.

[Dv7] G. David, Singular integrals on surfaces, Proc. Conf. on Harmonic Analysis and Partial Differential Equations, El Escorial 1987 (ed. J. García-Cuerva), p. 159-167, Lecture Notes in Math. 1384, Springer-Verlag 1989.

[DJe] G. David and D. Jerison, Lipschitz approximations to hypersurfaces, harmonic measure, and singular integrals, Indiana U. Math. Journal. 39, 3 (1990), 831-845.

[DJé] G. David and J-L. Journé, A boundedness criterion for generalized Calderón-Zygmund operators, Ann. of Math. 120 (1984) , 371-397.

[DJS] G. David, J-L. Journé and S. Semmes, Opérateurs de Calderón-Zygmund, fonctions para-accrétives et interpolation, Revista Matematica Iberoamericana, Vol 1, 4 (1985), 1-56.

[DS1] G. David and S. Semmes, L'opérateur défini par v.p. $\int \left| \frac{A(x)-A(y)}{x-y} \right| \frac{1}{x-y} f(y)\, dy$ est borné sur $L^2(\mathbb{R})$ lorsque A est lipschitzienne, C. R. Acad. Sci. Paris, t. 303, Série I, 11 (1986), 499-502 .

[DS2] G. David and S. Semmes, Strong A_∞ weights, Sobolev inequalities, and quasi-conformal mappings, Analysis and partial differential equations : a collection of papers dedicated to Mischa Cotlar, ed. C. Sadosky, Lecture notes in pure and applied mathematics 122, Marcel Dekker (1989) 101-111.

[DS3] G. David and S. Semmes, Singular integrals and rectifiabe sets in $\mathbb{R}^n$: au-delà des graphes lipschitziens, Astérisque 193, Société Mathématique de France 1991.

[DS4] G. David and S. Semmes, Quantitative rectifiability and Lipschitz mappings, to appear, Transactions A. M. S.

[DS5] G. David and S. Semmes, Harmonic analysis and the geometry of subsets of $\mathbb{R}^n$, Proceedings of the conference in honor of J-L. Rubio De Francia, El Escorial 1989.

[Do] J. R. Dorronsoro, A characterization of potential spaces, Proc. A.M.S. 95 (1985), 21-31.

[Fa] K. Falconer, The geometry of fractal sets, Cambridge University Press 1984.

[Fa1] K. Falconer, Fractal geometry, Mathematical foundation and applications, John Wiley & Sons, 1990.

[Fn] X. Fang, The Cauchy integral of Calderón and analytic capacity, Ph. D. Dissertation, Yale University 1990.

[Fe] H. Federer, Geometric measure theory, Grundlehren der Mathematishen Wissenschaften 153, Springer-Verlag 1963.

[FJ1] M. Frazier and B. Jawerth, Decomposition of Besov spaces, Indiana Univ. Math. J. 34 (1985), 777-799.

[FJ2] M. Frazier and B. Jawerth, The φ-transform and applications to distribution spaces, Function spaces and applications, M. Cwikel et al. ed., Lecture Notes in Math. 1302, Springer-Verlag 1988.

[FJ3] M. Frazier and B. Jawerth, A discrete transform and decompositions of distribution spaces, J. Func. Anal. 93, 1 (1990), 34-170.

[FJ4] M. Frazier, B. Jawerth, Applications of the φ and wavelets transforms to the theory of function spaces, to appear in Wavelets and their Applications, Coifman et al. ed., Jones and Bartlett, Boston.

[FHJW] M. Frazier, B. Jawerth, Y.-S. Han and G. Weiss, The T1-theorem for Triebel-Lizorkin spaces, Proc. Conf. on Harmonic Analysis and Partial Differential Equations, El Escorial 1987 (ed. J. García-Cuerva), p.168-181, Lecture Notes in Math. 1384, Springer-Verlag 1989.

[FJW] M. Frazier, B. Jawerth and G. Weiss, Littlewood-Paley theory and the study of Function spaces, NSF-CBMS Regional Conference series in Mathematics 79, AMS 1991.

[FK] M. Frazier and A. Kumar, The discrete wavelet trensform : an introduction, to appear in Wavelets : Mathematics and Applications, Benedetto and Frazier ed., CRC Press.

[FTW] M. Frazier, R. Torres and G. Weiss, The boundedness of Calderón-Zygmund operators on the spaces $F_p^{\alpha,q}$, Revista Matematica Iberoamericana, Vol. 4, 1 (1988), 41-72.

[GR] J. Garcia-Cuerva and J-L. Rubio De Francia, Weighted norm inequalities and related topics, North Holland, Amsterdam 1985.

[Ga1] J. Garnett, Positive length but zero analytic capacity, Proceedings A.M.S. 21 (1970), 696-699.

[Ga2] J. Garnett, Analytic capacity and measure, Lecture Notes in Math. 297, Springer-Verlag 1972.

[Ga3] J. Garnett, Bounded analytic functions, Academic Press 1981.

[GJ] J. Garnett and P. Jones, The corona theorem for Denjoy domains, Acta Math. 155 (1985), 29-40.

[Gr] K. Grochenig, Analyse multiéchelles et bases d'ondelettes, C. R. Acad. Sci. Paris, t. 305, Série I (1987), 13-17.

[Ge] F. Gehring, The L^p-integrability of partial derivatives of a quasiconformal mapping, Acta Math. 130 (1973), 265-277.

[Gu] M. De Guzman, Differentiation of integrals in $\mathbb{R}^n$, Lecture Notes in Mathematics 481, Springer-Verlag 1975.

[HT] M. Holschneider and P. Tchamitchian, Pointwise analysis of Riemann's "non differentiable" function, Inventiones Mathematicae 105 (1991), 157-176.

[Ja] S. Jaffard, Construction et propriétés des bases d'ondelettes ; remarques sur la contrôlabilité exacte, Doctorat, Ecole Polytechnique.

[Ja2] S. Jaffard, Pointwise smoothness, two-microlocalization and wavelets coefficients, Publicacions Matemàtiques (Universidad Autonòma de Barcelona) vol 35 (1991), 155-168.

[JaM1] S. Jaffard and Y. Meyer, Bases d'ondelettes dans les ouverts de $\mathbb{R}^n$, Journ. Math. Pures et Appl. 68 (1989), 95-108.

[JaM2] S. Jaffard and Y. Meyer, Les ondelettes, Proc. Conf. on Harmonic Analysis and Partial Differential Equations, El Escorial 1987 (ed. J. García-Cuerva), p. 182-192, Lecture Notes in Math. 1384, Springer-Verlag 1989.

[JK] D. Jerison and C. Kenig, Boundary behaviour of harmonic functions in non-tangential access domains, Adv. in Math. 46 (1982), 80-147.

[Jn1] P. Jones, Square functions, Cauchy integrals, analytic capacity, and harmonic measure, Proc. Conf. on Harmonic Analysis and Partial Differential Equations, El Escorial 1987 (ed. J. García-Cuerva), p. 24-68, Lecture Notes in Math. 1384, Springer-Verlag 1989.

[Jn2] P. Jones, Lipschitz and bi-lipschitz functions, Revista Matematica Iberoamericana, vol.4, 1 (1988), 115-122.

[Jn3] P. Jones, Rectifiable sets and the traveling saleseman problem, Inventiones Mathematicae 102, 1 (1990), 1-16.

[JnM] P. Jones and T. Murai, Positive analytic capacity but zero Buffon needle probability, Pacific J. Math. 133 (1988), 99-114.

[Jé] J-L. Journé, Calderón-Zygmund operators, pseudodifferential operators and the Cauchy integral of Calderón, Lecture Notes in Math. 994, Springer-Verlag 1983.

[Le1] P-G. Lemarié, Continuité sur les espaces de Besov des opérateurs définis par des intégrales singulières, Ann. Inst. Fourier Grenoble 35 (1985), 175-187.

[Le2] P-G. Lemarié, Une nouvelle base inconditionnelle de $H^1(\mathbb{R}\times\mathbb{R})$, preprint.

[Le3] P-G. Lemarié, Fonctions à support compact dans les analyses multi-résolutions, Revista Matematica Iberoamericana, Vol. 7 (1991), 157-182.

[LM] P-G. Lemarié and Y. Meyer, Ondelettes et bases Hilbertiennes, Revista Matematica Iberoamericana, Vol 2, 1(1986), 1-18.

[Ma] S. Mallat, Multiresolution approximation and wavelet orthonormal bases of $L^2(\mathbb{R})$, Transactions A. M. S. 315 (1989), 69-87.

[McM] A. McIntosh and Y. Meyer, Algèbre d'opérateurs définis par des intégrales singulières, C.R. Acad. Sci. Paris, t. 301, Série I, 8 (1985), 395-397.

[McS1] R. A. Macias and C. Segovia, Lipschitz functions on spaces of homogeneous type, Adv. in Math. 33 (1979), 257-270.

[McS2] R. A. Macias and C. Segovia, A decomposition into atoms of distributions on spaces of homogeneous type, Adv. in Math. 33 (1979), 271-309.

[MRV] O. Martio, S. Rickman et J. Väisälä, Definitions for quasiregular mappings, Ann. Acad. Sci. Fenn., Ser. AI, 448 (1969), 1-40 .

[Mt] P. Mattila, Lecture notes on geometric measure theory, Universidad de Extramadura, Spain (1986).

[Mt2] P. Mattila, A class of sets with positive length and zero analytic capacity, Ann. Acad. Sci. Fen., 10 (1985), 387-395.

[Mt3] P. Mattila, Smooth maps, null-sets for integralgeometric measure and analytic capacity, Ann. of Math. 123 (1986), 303-309.

[Mt4] P. Mattila, Cauchy singular integrals and rectifiability of measures in the plane, preprint.

[My1] Y. Meyer, Les nouveaux opérateurs de Calderón-Zygmund, Colloque en l'honneur de Laurent Schwartz, Vol. 1, Astérisque 131 (1985), 237-254.

[My2] Y. Meyer, Principe d'incertitude, bases hilbertiennes et algèbres d'opérateurs, Séminaire Bourbaki, 1985-86, 662, Astérisque (Société Mathématique de France).

[My3] Y. Meyer, Constructions de bases orthonormées d'ondelettes, Revista Matematica Iberoamericana, Vol. 4 (1988), 31-40.

[My4] Y. Meyer, Ondelettes et Opérateurs (3 volumes), Herman 1990.

[My5] Y. Meyer, Ondelettes sur l'intervalle, Revista Matematica Iberoamericana, Vol. 7 (1991), 115-134.

[My6] Y. Meyer, Ondelettes et applications, Notes de cours d'une série d'exposés à l'Institut d'Espagne en 1991, Univ. Paris-Dauphine (CEREMADE). Will also exist in spanish.

[Mo] F. Morgan, Geometric measure theory, A beginner's guide, Academic Press 1988.

[Mu1] T. Murai, Boundedness of singular integral operators of Calderón type VI, Nagoya Math. Journal 102 (1986), 127-133.

[Mu2] T. Murai, A real variable method for the Cauchy transform and analytic capacity, Lecture Notes in Math. 1307, Springer-Verlag 1988.

[Mu3] T. Murai, The power 3/2 appearing in the estimate of analytic capacity, Pacific Journal of Math. 143 (1990), 313-340.

[Ond1] Wavelets : time-frequency methods and phase space, Proceedings int. conf. Marseille 1987, Combes, Grossman and Tchamitchian ed., Springer-Verlag 1989 (IPTI).

[Ond2] P.-G. Lemarié ed., Les Ondelettes en 1989, harmonic analysis seminar at Orsay, L.N. in Math 1438, Springer-Verlag 1990.

[Ru] J-L. Rubio de Francia, Factorization theory and A_p weights, Am. J. Math. 106 (1984), 533-547.

[Se1] S. Semmes, The Cauchy integral, chord-arc curves, and quasiconformal mappings, proceedings of the symposium on the occasion of the proof of the Bieberbach conjecture, A.M.S. Math. Surveys 21, 1986,167-184.

[Se2] S. Semmes, Nonlinear Fourier analysis, Bulletin A. M. S. 20 (1989), 1, 1-18.

[Se3] S. Semmes, Chord-arc surfaces with small constants I, to be published, Advances in Math.

[Se4] S. Semmes, Chord-arc surfaces with small constants II: good parametrizations, to be published, Advances in Math.

[Se5] S. Semmes, A criterion for the boundedness of singular integrals on hypersurfaces, Trans. A.M.S. 311, 2 (1989), 501-513.

[Se6] S. Semmes, Square function estimates and the T(b) theorem, to be published, Proc. A.M.S.

[Se7] S. Semmes, Hypersurfaces in $\mathbb{R}^n$ whose unit normal has small BMO norm, preprint.

[Se8] S. Semmes, Differentiable function theory on hypersurfaces in $\mathbb{R}^n$ (without bounds on their smoothness), Indiana Univ. Math. Journal 39 (1990), 985-1004.

[Se9] S. Semmes, Analysis vs. geometry on a class of rectifiable hypersurfaces in $\mathbb{R}^n$, Indiana Univ. Math. Journal 39 (1990), 1005-1036.

[Se10] S. Semmes, Bilipschitz embeddings of $\mathbb{R}^n$ into $\mathbb{R}^N$ for A_1-weights, preprint

[St] E. M. Stein, Singular integrals and differentiability properties of functions, Princeton University Press 1970 .

[St2] E. M. Stein, Topics in harmonic analysis related to the Littlewood-Paley theory, Princeton University Press 1970.

[Sg] O. Stromberg, A modified Franklin system and higher-order systems of $\mathbb{R}^n$ as unconditional bases for Hardy spaces, Conference on harmonic analysis in honor of Antoni Zygmund, Vol. II, 475-493, ed. W.Beckner and al., Wadworth math. series 1983.

[Tc1] P. Tchamitchian, Calcul symbolique sur les opérateurs de Calderón-Zygmund et bases inconditionelles de $L^2(\mathbb{R})$, C. R. Acad. Sci. Paris, t. 303, Série I, 6 (1986), 215-217.

[Tc2] P. Tchamitchian, Biorthogonalité et théorie des opérateurs, Revista Matematica Iberoamericana vol.3, 2 (1987), 163-190.

[Tc3] P. Tchamitchian, Ondelettes et intégrale de Cauchy sur les courbes lipschitziennes, Ann. of Math. 129 (1989), 641-649.

[Tc4] P. Tchamitchian, Inversion des opérateurs de Calderón-Zygmund, preprint.

[Tr] H. Triebel, Theory of function spaces, Monographs in Mathematics, Vol. 78, Birkauser, Basel, 1983.

[Vä1] J. Väisälä, Uniform domains, Tôhoku Math. J. 40 (1988), 101-118.

[Vä2] J. Väisälä, Invariants for quasisymmetric, quasimöbius and bilipschitz maps, J. d'Analyse Math. 50 (1988), 201-233.

[W] M. V. Wickerhauser, Acoustic signal compression with Walsh-type wavelets, preprint.

[Z] A. Zygmund, Trigonometric series, Cambridge University Press 1968.

NOTATIONS

INDEX